DATE DUE

OCT 9 '84			

American Farmers

Minorities in Modern America

Editors
Warren F. Kimball
David Edwin Harrell, Jr.

American Farmers

The New Minority

Gilbert C. Fite

INDIANA UNIVERSITY PRESS
Bloomington

For the four Furrys–Bill,
Wanda, Carl, and Evelyn–
good farmers and good friends.

Manufactured in the United States of America

Library of Congress Cataloging in Publication Data

Fite, Gilbert Courtland, 1918—
American farmers.

(Minorities in modern America)
Includes index.
1. Farmers—United States. 2. Farmers—United States—Political activity. 3. Agriculture—Economic aspects—United States. 4. Agriculture and state—United States. 5. United States—Rural conditions. I. Title. II. Series.
HD8039.F32U64 306'.3 80-8843
ISBN 0-253-30182-3 AACR2
1 2 3 4 5 85 84 83 82 81

Contents

Preface

Of all the minorities in American society, farmers are the only group that once held a clear, even an overwhelming, majority. The transition from majority to minority occurred gradually as the economy became more industrialized and people moved to urban jobs. After about 1940, however, the shift out of farming became dramatic. Technological and scientific advances in agriculture reduced the need for agricultural labor, and industrial expansion absorbed millions of unneeded farmers. The declining number of farm people reflected basic changes in the economy of the United States—agricultural to industrial, and rural to urban.

My main attention in this book has been directed toward the problems and progress of family size, commercial farmers who have been the principal producers of the nation's food and fiber. In seeking to explain the forces that have reduced this important group of Americans to such a small number, I have discussed the technological, economic, and social developments that have drastically affected agriculture and life on the farm since the 1920s. I have concentrated on the economic and political position of farmers and on how they responded to their declining influence in American society. This has required substantial discussion of agricultural policy and farm politics. Because the cost-price squeeze has forced so many farmers out of agriculture, I have dealt with this question in some detail. The ideas and emotions associated with the agrarian tradition, and the concept of the family farm, have been woven through this study to show how deeply Americans have been affected by the value systems associated with agriculture and farm life.

Readers are entitled to know my general views and assumptions. Raised on a South Dakota farm which I still own, I admit to a definite rural perspective. I hold strong and favorable feelings about land, agriculture, and farm people. However, my rural sympathies have been tempered by the fact that life was frightfully hard for many farmers, especially before World War II. I hold no brief for the so-called good old days down on the farm, simply because experience and statistics both confirm that they were not very good for millions of American farmers. I find little fault with the developments in commercial agriculture. If farmers were successfully to keep up with, and participate in, the modernizing of America, they had to employ science and technology in their operations and emphasize the business aspects of farming. Otherwise, in most cases, they could not make a living comparable to that earned in other employments. I reject policies that deny farmers the possibility of making enough money for a middle-class standard of living, and do not believe farmers should be expected to produce food and raw materials at unprofitable prices for the benefit of other and more powerful elements in American society.

This study has not been written for specialists in agriculture, economics, politics, or agricultural history, although it may be of some interest to people in all of these disciplines. It is presented to the general reader who wants to view the broad outline of change surrounding farming and farm life since about 1920. A conscious effort has been made to let farmers speak to the issues and problems when that seemed appropriate. Chapter twelve deals entirely with the views and opinions of modern farmers on a wide variety of questions and issues relating to their welfare.

I have had much assistance in the preparation of this book. The University of Georgia has been most generous in providing time for research and writing. The efficient work of my research assistants, David D. Potenziani and Edward C. Nagy, saved me many hours of tiring effort and hastened completion of the study. Librarians and curators of manuscript collections have also helped locate materials that I might otherwise have missed. Roe C. Black, executive editor of *Farm Journal*, assisted me in surveying farm attitudes and also gave me the benefit of his wide knowledge of agriculture and its problems. Scores of farmers and farm leaders have taken time to visit with me and provide information that simply is not in the written record. In this regard, I am especially grateful to William Furry of Charleston, Illinois, and Robert Black of Algona, Iowa, who shared their ideas and knowledge that grew out of their long experience as successful commercial family farmers.

For the pictures I am indebted to the United States Department of Agriculture, Farmland Industries, Inc., the Tennessee Valley Authority, the University of Georgia College of Agriculture, and Verlyn Larson.

To my wife, June, who dislikes everything about the farm she left many years ago, I express my appreciation for tolerating my sorties into agricultural history, a subject in which she has no interest whatever. Finally, I am grateful to Mrs. Edna Fisher, my secretary, for her loyalty and efficiency.

Gilbert C. Fite
Athens, Georgia

American Farmers

The Vanishing Majority

IT WOULD HAVE BEEN INCONCEIVABLE IN THE EARLY DAYS of the Republic for anyone to imagine a time when farmers would make up only a small minority of the nation's population, and that agriculture would fall from primary importance in the economy. The first United States census, taken in 1790, found that at least 96 percent of the people were rural and that most of them actually lived on their individual farms. The new nation was a country of farmers and planters, and agriculture generated most of the nation's income. Moreover, the vast majority of Americans subscribed to the view that farming represented a superior way of life. As Thomas Jefferson wrote in 1781, "those who labour in the earth are the chosen people of God, if ever he had a chosen people." President George Washington told Congress in 1796 that "it will not be doubted that with reference either to individual or national welfare agriculture is of primary importance."

As the nineteenth century opened, a vast and potentially rich agricultural domain between the Appalachians and the Mississippi River lay open for settlement by land-hungry and energetic pioneers. Except for the Indians, whose claims to land were seldom recognized, there was enough land for hundreds of thousands of new farms. Then in 1803, the United States acquired the Louisiana Purchase—529 million acres—nearly doubling the nation's size. Within another half century, either through treaty or by war, Florida, Texas, Oregon, and the desert Southwest between Texas and the Pacific were brought within America's boundaries. Altogether the continental United States finally extended over 3,022,387 square miles, or nearly 2 billion acres. So rich, varied, and extensive were the lands within the United States that they appeared inexhaustible. Jefferson believed that there was room "enough for our descendants to the thousandth and thousandth generation."

Most of the western lands were in the public domain. This meant that federal policy played a significant role in how the land would be distributed and settled. Despite a spirited controversy in Congress over whether land should be sold to produce revenue or distributed on easy terms to settlers, the federal government gradually sided with the farmer. Beginning in 1800, Congress passed a series of laws to make it easier for farmers to acquire land directly from the federal government. The law of 1820 permitted a person to buy as little as 80 acres directly from the government for $1.25 an acre. A farmer could now obtain a farm for as little as $100. Speculators also took advantage of the land laws, but land was so abundant that they could not monopolize it or deny a farm to ordinary settlers.[1]

The prospect of acquiring a piece of land, a share of the national patrimony, attracted swarms of settlers to the unoccupied West. Following the War of 1812, thousands of farmers and planters rushed into the new lands of the Old Northwest and the Old Southwest. Families moved into Ohio, Indiana, Illinois, and even across the Mississippi River into Missouri. The Englishman Morris Birkbeck described the rush into the Ohio Valley in 1817: "Old America seems to be breaking up, and moving westward. We are seldom out of sight . . . of family groups, behind and before us." At the same time thousands of small farms and large plantations were established in Alabama, Mississippi, Arkansas, and Louisiana.

Northern farmers found the midwestern soil and climate ideal for wheat, corn, and other grains, as well as livestock. In the South, medium and large planters tended to concentrate on the great commercial crops of cotton, tobacco, rice, and sugar, along with some food products. The yeoman farmers raised corn and other grains, livestock, and small amounts of cotton and tobacco. Elimination of the Indian menace, together with federal land policies, good soil, better transportation, a growing demand for American products (especially cotton) both at home and abroad, improved farming practices, more and better tools and machines, all combined to expand the number of farms and to increase agricultural productivity in the half century before the Civil War.[2]

Wherever one looked on the American scene in the early nineteenth century the importance and predominance of farming was clearly evident. American food may not have been fancy, but it was abundant for most people. The developing factories, such as those for woolen and cotton textile industries, meat-packing, and flour milling, all depended on farm-produced raw materials. Most of the cargoes on wagons, boats, ships, and later the railroads consisted of farm products.

For a developing country, exports to earn foreign exchange are highly important. It was farmers who produced the vast majority of commodities shipped abroad in the nineteenth century. Exports of cotton, tobacco, grain, and meat brought millions of dollars to the United States. In 1860, cotton exports alone amounted to $192 million. Douglass C. North, a leading economic historian, credits cotton exports with contributing mightily to the nation's economic development, especially in the years from 1823 to 1843. He refers to cotton as the "proximate prime mover in quickening the pace of the country's [economic] growth."[3] By 1860, about 80 percent of the total exports of the United States came from the nation's farms, ranches, and plantations.

Most American farmers were to some degree self-sufficient. That is, they produced a portion of their own needs on the farm. This included food, furniture, tools and utensils, and clothing. But farmers moved quickly into commercial operations during the early nineteenth century, and self-sufficiency declined rapidly. It became more and more common for producers to sell their surplus commodities and to purchase household articles or items needed in the operation of the farm. The large planters who employed slave labor were the most highly commercialized operators, but even small, relatively self-sufficient farmers produced grain, tobacco, pork, and other products for sale.

Despite growing industrialization and urbanization in the first half of the nineteenth century, agriculture continued to be the nation's largest business and farmers the most numerous group of workers. Before 1920, the Census Bureau did not separate the population actually living on farms from those residing in small towns and villages. The rural population, defined as those living on farms and in towns of less than 2,500, slightly exceeded 80 percent of the total population in 1860. Of 31.4 million people in the United States, 25.2 million were either on farms or in small towns. Most of these were farmers and their families.

In 1860 there were 2,044,077 farms, an increase of 595,004 during the previous decade. Although American farms varied from small plots to large plantations and ranches of thousands of acres, the average size was 199 acres. The value of farm land and buildings amounted to $6.6 billion. This was more than six times the value of the 139,722 manufacturing establishments. Farming was responsible for more than 30 percent of the national income and for nearly 60 percent of all employment. On the eve of the Civil War, there were few indications that farmers would not maintain their predominant position in the nation's economy and culture for generations to come.

The idea that farmers were God's chosen people and that farming was man's most useful occupation continued strong in American thinking. This agrarianism or agriculture fundamentalism held that farm life produced better people and that citizens close to the soil were more democratic, honest, independent, virtuous, self-reliant, and politically stable than were city dwellers. In farming, it was said, man worked hand in hand with the Creator to supply people's physical needs.

Praise of farmers and farming came from presidents, editors, politicians, poets, and other opinion-makers. In 1832, in his fourth annual message to Congress, President Andrew Jackson declared: "The wealth and strength of a country are its population, and the best part of that population are cultivators of the soil. Independent farmers are everywhere the basis of society and the true friends of liberty." Ralph Waldo Emerson wrote that agriculture "has in all eyes its ancient charm, as standing nearest to God, the first cause."[4]

As might be expected, editors of the growing number of farm journals were even more enthusiastic in their devotion to agriculture and rural life. Jesse Buel, founder and editor of *The Cultivator*, wrote in 1836 that "every business in life is mainly dependent, for its prosperity, upon the labors of agriculture." He added that other classes "cannot thrive . . . without the aid of the farmer: he furnishes the raw materials for the manufacturer, he feeds the mechanic, and freights the bark of commerce; and is besides the principal customer of them all." Writing to the editor of the *Southern Cultivator* in July 1844, one reader waxed even more eloquent when he declared: "the farmer is the main support of human existence. He is the lifeblood of the body politic, in peace and war, . . . freedom, patriotism and virtue, after being driven from the degeneracy and corruption of the cities, will find their last resting place in the bosom of the agriculturalist."

With the economic dominance of agriculture and wide support for agrarian ideals, it should have followed that farming interests would have been the special concern of lawmakers. Logic would dictate that in a democracy farm welfare should have been adequately protected and fostered through the influence of sheer numbers. However, this was not the case.

Speaking on this subject before the New York State Agricultural Society in 1856, Samuel Cheever, the society's president, declared that "in the sharp struggles for special advantages between different interests in our country, of which our National and State Legislatures have frequently been the scenes," the agricultural interest, "preeminent as it should be, has been too often overlooked or disregarded." Another pre–Civil War observer remarked that "the commercial and manufacturing interests, being locally limited and centralized, can easily combine and make themselves felt in the halls of legislation, and in the executive departments of the government. Not a session of congress passes without this being clearly and sometimes painfully evident. New York and Lowell have often more immediate influence in directing and molding national legislation than all the farming interests in the country. Agriculture, clad in homespun, is very apt to be elbowed aside by capital, attired in ten-dollar Yorkshires." The main reasons politicians paid relatively little heed to agricultural interests were that farmers were not organized or united, they were widely scattered, and their interests often conflicted. As Cheever saw it, "united action" and "associated wealth" were important in political influence. These, he said, "farmers possess less than any other class."

Except for those who favored higher tariffs on certain farm products, and land legislation, Congress paid little attention to particular farm needs in the pre–Civil War years. After complaining in 1845 that the southern states were not doing anything for farmers, the editor of the *Southern Cultivator* said that farmers would be ignored "until the tillers of the soil assert their rights to an equal participation in the benefits of government." Speaking about the tariff, President James K. Polk told Congress in 1846 that even though farmers constituted "a large majority of our population," they "have heretofore not only received none of the bounties or favors of Government" but "had endured burdens that enriched others." Five years later President Millard Fillmore told Congress that "justice and sound policy" required the federal government to utilize all constitutional means "to promote the interests and welfare" of farmers.

Lack of congressional concern for agricultural interests may have been related to the fact that farmers were poorly represented in the halls of Congress. They never had representation proportionate to their numbers. From 1789 to 1800, a time when approximately 90 percent of the population was farmers, only 12.9 percent of the members of the House of Representatives were farmers or connected with agriculture. By the 1840s the figure had declined to 8.6 percent.[5]

In 1820 the House of Representatives established a Committee on Agriculture to deal with farm interests in Washington. In pushing for his resolution, Congressman Lewis Williams of North Carolina said: "And how happens it, sir, that the agricultural, the great leading and substantial interest in this country, has no committee—no organized tribunal in this House to hear and determine their grievances?" Five years later the Senate created its Committee on Agriculture and Forestry. Congressional machinery was at last available to give special attention to farm concerns.[6]

Federal lawmakers made their first direct appropriation for agriculture in 1839, when they provided $1,000 to gather agricultural statistics and to distribute free seeds for experimental purposes. The first agricultural census was also taken that year. During the next twenty years, however, farmers received little attention from Congress. In 1851, in his annual message to Congress, President Fillmore observed that "the manufacturing and commercial interests have engaged the attention of congress during a large portion of every session and our statutes abound in provisions for their protection and encouragement." Yet, he added, "little has yet been done directly for the advancement of agriculture." He recommended establishing an agricultural bureau that would collect and disseminate information on the best methods of cultivation, distribute seeds, and show farmers how to conserve and restore the soil and how best to grow certain crops. But demands by presidents, congressmen, editors, and farm spokesmen for the creation of a bureau or department of agriculture went unheeded for more than another decade. The little assistance directed to agriculture during the pre–Civil War years was provided by a representative in the Patent Office.

Secession of the South in 1861 removed the sectional conflicts in Congress and opened the way for passage of three highly significant pieces of legislation beneficial to farmers. In 1862, following years of agitation, Congress passed the Homestead Act, established a Department of Agriculture, and provided federal support for the creation of agricultural and mechanical colleges. The Homestead Act offered, at no cost except for a small filing fee, 160 acres of land to every qualified citizen who agreed to live on it for a specified time. The Department of Agriculture, headed by a commissioner, continued to distribute seeds and to provide agricultural information, but it also soon became quite heavily involved in plant and animal research. The department received cabinet rank in 1889, at which time it had 488 employees and a yearly appropriation of $1.1 million. In order to facilitate practical agricultural education, Congress passed the Morrill Act in 1862, which gave 30,000 acres of public land for each senator and representative in a state (or land certificates or "script" in states where land was not available) to support the establishment of an agricultural and mechanical college. These "land grant" institutions, along with federally supported agricultural experiment stations begun in 1887 with passage of the

Hatch Act, did much to expand agricultural research and to promote practical education among farmers.[7]

In the half century following the Civil War unprecedented agricultural expansion occurred. Americans occupied more land and established a greater number of farms, mostly west of the Mississippi River, than in any comparable time in history. Between 1860 and 1910 the amount of land turned into farms exceeded that of the preceding 250 years. Farm acreage in that half century jumped from 407 million to 879 million acres. The number of farms increased more than three times, rising from just over 2 million in 1860 to almost 6.4 million in 1910. In the 1870s alone nearly 1.4 million new farms were established, and nearly that many were added in the 1890s.

American farmers produced a flood of cotton, tobacco, corn, wheat, and other grains, as well as millions of head of livestock and a variety of specialty crops. Production far exceeded domestic demand, and cotton, wheat, tobacco, pork, beef, and other products flowed to world markets. Exports of cotton reached 4.4 million bales in 1880–81 and jumped to 6.6 million bales in the 1900–1901 season. The amount of wheat sold abroad exceeded 200 million bushels annually by 1900, and in that year meat sales overseas rose above $100 million. Total American exports amounted to $836 million in 1880, of which about 70 percent was farm products. In the first years of the twentieth century more than 60 percent of the nation's exports still came from the farms. Furthermore, two of the country's four leading industrial groups—food and kindred products, and textiles—depended on the farms for their raw materials.[8]

A variety of favorable factors, already at work before the Civil War, combined to produce this abundance. Good soil and favorable climate in the regions where many new farms were established contributed greatly to agricultural productivity. Moreover, liberal federal land laws helped thousands of farmers get established. More and better farm machinery, improved farming methods, better management, and the creation of a national railroad network were among other developments that improved farm efficiency and productivity.

The great increase in the number of farms in the late nineteenth century as settlers pushed westward to the Pacific demonstrated a continued strong demand for land and economic opportunities in agriculture. The nation's rural population increased substantially, rising from approximately 18 million in 1850 to 35.8 million in 1880 and to 49.8 million by 1910.

Despite the phenomenal growth of agriculture following the Civil

War, the predominant position of farmers and farming was rapidly ebbing away. For a time the transformation being brought about by surging industrialism was barely perceptible, but every decennial census brought the changing situation into clearer focus. While agriculture's absolute gains were impressive, by whatever criteria applied—population growth, number of farms, employment, or production of wealth—the relative position of farming showed a dramatic decline.

For example, the rural population rose by 14 million between 1880 and 1910, but the number of city dwellers increased by some 28 million, or twice as fast. In the last decade of the nineteenth century the urban population grew at a rate of 36 percent compared to only 12 percent on farms and in small villages. Migration cityward in the Northeast had been strong even before the Civil War, but it became an avalanche after 1860. New York City had 1.2 million people in 1860, but contained 4.7 million a half century later as both native-born Americans and foreign immigrants flooded into the area. Boston, Baltimore, Philadelphia, and scores of smaller cities also showed major gains. Urbanization, however, was not confined to the Northeast, where commercial and industrial activity first developed on a large scale. Midwestern and western cities also experienced remarkable gains. Chicago grew from only 109,260 to 2.2 million between 1860 and 1910. The Windy City increased 118 percent in the expansive 1880s. While urbanization lagged in the South, even there significant urban development was taking place by the end of the nineteenth century. In the entire nation, only about 22 percent of the people were classified as urban in 1860; by 1900 it was 39 percent; and the census of 1920 showed that for the first time in American history over half—51.4 percent—of Americans were urban. The phenomenal growth of cities and the relative decline of agriculture made America an urban nation by the early twentieth century, reversing the condition that extended back to the colonial era and the early days of the Republic.

As the population shifted from rural to urban, the labor force changed. In 1860, for example, farm workers made up about 60 percent of the nation's labor force. This figure dropped to 42.7 percent by 1890 and to only 30.5 percent by 1910. There were nearly as many people employed in manufacturing and mechanical pursuits in 1910 as there were in agriculture. The basic shift in the economy can also be seen in the production of wealth. Historically, farming had been responsible for the largest share of the nation's wealth and income. In 1799 farming was responsible for nearly 40 percent of private production income, and it was still some 31 percent in 1859.

By 1900, however, agriculture produced only 20 percent of the nation's income. The census of 1890 showed that for the first time the value of industrial output exceeded the wealth produced on farms. There were some years after that when the value of farm products was greater than that coming from the nation's factories, but the tide had turned. A fundamental and far-reaching change had occurred in American society. The farm majority and the predominance of agriculture were at an end. The eighteenth and nineteenth centuries belonged to the American farmer; the twentieth was being claimed by business and industry.

Statisticians saw the situation clearly, but it took many years for the general public to grasp the fundamental change that had occurred. Popular consciousness always lags behind actual social and economic change. Furthermore, farmers continued to *appear* much more important in the overall life of the nation than they actually were. The number of farmers and small-town residents continued to grow, at least until 1910, and they made up a major group in every part of the country except the Northeast and the West Coast. Even though Chicago had over 2 million persons by 1900, the Midwest, of which Chicago was a part, continued to be mainly rural. A train trip through the heartland of America—from Pittsburgh to Chicago to Kansas City and on to Denver—at the turn of the century would have given every impression that the farmer was king. Travel throughout the South would have strengthened that view even more.

Long before farmers became a minority, many citizens expressed alarm at the way cities and urban employment were pulling people away from the countryside. As early as the 1820s and 1830s, abandoned farms dotted New England and parts of the Middle Atlantic states as manufacturing and nonfarm employment opened up job opportunities for rural residents. The West also beckoned to many farm people in the Northeast. Describing the move toward cities, a writer for the Windsor *Vermont Chronicle* reported in 1845 that "Within a few weeks, the daughters of Vermont have passed our doors by the score at a time, to be employed in factory work in another state."

The movement from farm to factory accelerated throughout the nineteenth century. Continued praise of farm life by politicians, writers, leaders of agricultural societies, and others could not change the trend. D. J. Baker told the Illinois Agricultural Society at the State Fair in 1855 that "the culture of the earth is the only trade which man has ever been commanded by his maker to exercise." "Agricultural pursuits," he continued, "are the handmaids to innocence, peace and self-satisfaction . . . and are among the most honorable employment in which we can be engaged." Speaking at the first territorial fair in Dakota in 1885, Governor Gilbert A. Pierce declared that "this nation must depend largely upon its agricultural population to preserve its honor, its integrity, and its liberty."

To many Americans the cityward movement represented growing national degeneracy and a weakening of the Republic. If young citizens would not respond to the idea that farming was the best way to make a living and the most pleasant life, agricultural fundamentalists sought to discourage migration by picturing growing urban centers as sinkholes of sin, crime, and corruption. "All great cities," said a speaker at Bennington, Vermont, in 1857, "are cursed with immense accumulations of ignorance and error, vice and crime and misery; . . . their theaters and gambling houses, drunkeries and brothels are well patronized; . . . Far better is it for our youth to breathe the pure air and enjoy the salutary moral influences of their native state, than to be brought into contact with such masses of putrefaction."

Many of the most ardent supporters of agriculture and farm life, as well as those highly critical of cities, were not farmers. The strongest proponents of the agrarian faith were usually people who had left the farm and made a living in some nonfarm employment. Writing in the *Southern Cultivator* in 1846, one correspondent saw the situation clearly: "Unfortunately for agriculture," he wrote, "its loudest and most conspicuous admirers are constantly lavishing upon it expressions of respect, while, at the same time, they disdain the idea of proving their sincerity by any act whatever. They admire the profession but advise their sons to pursue another." N. S. Hubbard told a Massachusetts audience in 1872 that "we find . . . almost invariably that those who speak in the highest terms of agriculture are those who do not get their income exclusively from the farm."[9]

Here was a strange contradiction. Agrarian rhetoric and theory held that agriculture was the nation's basic industry and that farming represented a superior way of life. Farmers were freer, more independent, self-sufficient, honest, dependable, and devoted to high moral principles. Yet millions of Americans were deserting this idyllic condition, leaving the fields of their fathers, and flooding to an industrial and urban environment reputedly filled with corruption, temptation, and vice. However, the contradiction was only in theory. The reality was that agriculture could not compete with industry and urban life, a fact that became increasingly evident as the nineteenth century advanced.

Why was this the case? What caused the rural exodus that threatened the agricultural majority? The answers to these questions varied greatly among the individuals involved, but several major reasons are clear. In the first place, except for a few large planters and ranchers, farmers suffered from low social status. Compared to business, the professions, or even factory work, agriculture ranked low. The "hick" and "hayseed" concept developed early. Anyone, it was said, could farm. One observer wrote in the early 1840s that there was a tradition that if a family had a crippled child he would become a tailor or a minister; if here were a blockhead or a dunce he would take up farming. Farming was not viewed as a profession, but as a round of drudgery and monotony which youth sought to avoid. As one New Englander complained in the 1830s, "Every farmer's son and daughter are in pursuit of some genteel mode of living. After consuming the farm in the expenses of a fashionable, flashy, fanciful education, they leave the honorable profession of their fathers to become doctors, lawyers, merchants, or ministers or something of the kind."[10]

Although the increased use of horse-drawn machinery after about 1850 helped to reduce the physical demands of farming, agricultural employment continued to be characterized by hard work. Throughout much of the year a farmer's efforts were governed by the seasons, the crops, and the demands of livestock. He and his family worked from early morning until late at night. Complaints about hard work and long hours on the farm were abundant. As one observer wrote in 1896, much of farm life was "drudge, drudge, from daylight to dark, day after day, month after month, year after year." Whatever the sentimentalists may have written about the noble calling of agriculture and its elevating and purifying influences, farm youth were not, as one commentator expressed it, "caught up in this sort of chaff." To a farm boy, he continued, "Mother Earth is an exacting parent, calling for constant and regular toil, and whipping him on day by day with weeds to be hoed, dry gardens to be watered, . . . and an almost endless round of embarrassments to be overcome." Testifying before the Industrial Commission in 1899, one witness said that young men and women "tend to consider farm labor unpleasant and dirty." Sam Rayburn, a young Texas congressman, declared in 1916 that "some of our city friends talk about the beautiful life on the farm," but that concept would vanish if those people "would go out and bend their backs over a cotton row for 10 or 12 hours [or] grip the plow handles that long."

Not only was farm labor hard and distasteful, according to many observers it was also dull and monotonous. Such expressions as "breaking prairie," "hoeing corn," and "plowed all day" were common phrases found in diaries kept by farmers in the nineteenth century. While there was wide variation of tasks required on the farm, they soon fell into a kind of monotonous regularity. A common example of this was the need to be present every morning and night to milk the cows, a task heartily detested by many farm youth.

Numerous commentators pointed to the isolation, lack of proper educational opportunities, and scarcity of social and cultural advantages as factors that made the countryside less desirable than the city. The rural life-style simply could not compete with the conveniences, cultural opportunities, social amenities, and entertainments that a concentrated population could provide. The editor of the Milwaukee *Sentinel* explained on October 29, 1889, that cities had good walks, lights, the best schools, churches, shows, bands, newspapers, running water, and, most important of all, people with whom to socialize.

Furthermore, farming, by its very nature, was risky. The undependable weather, animal and plant diseases, crop-destroying pests, and uncertain prices produced ceaseless worry and anxiety among farmers. They never knew at the beginning of the planting season how they would come out at harvest time. It is no wonder that for generations the main topics of conversation among farmers were the weather, crops, and prices. No amount of hard work and good management could remove the chance factors in agriculture. Writing to the *Prairie Farmer* in 1875, one correspondent said that the farm youth he had talked to unanimously condemned farming and favored an occupation with a regular salary. They preferred the security of a city job, even though they might not get rich, to the independence of the country "without any such chance." Referring to farmers leaving agriculture for textile mills in the South, a witness before the Industrial Commission said that factory work was not necessarily preferred, but workers got "cash pay every Saturday night."

The lack of challenge and intellectual stimulation caused many rural youth to forsake the farm. The editor of the Milwaukee *Sentinel* wrote on October 27, 1889, that "the farm had little to offer to an active young mind." E. J. Wickson, speaking at the University of California, said he believed that youth were leaving agriculture because "in the minds of the nobler class of young men it [the move from the farm] is a desire for improvement, an ambition for wider success, an impulse to greatness." Born on a Nebraska homestead in 1874, Alvin Johnson, a well-known economist, recalled that when he was a boy his mother had urged him to leave the farm and take up another occupation. He would wear himself out prematurely with hard work, she told him, by the time men in other professions would just be reaching their prime. Darwin P. Kingsley, a life insurance executive, explained that life was moral and self-respecting in the rural community of his boyhood, but "it was extremely narrow, uninspiring, and unimaginative. There was little or nothing to fire a boy with ambition or enthusiasm or to acquaint him with the world that lay beyond." The literature dealing with farm to city migration is filled with such testimony.

Throughout the nineteenth century there was widespread consideration of how to make farm life more attractive in order to keep rural youth from rushing off to the towns and cities. This very discussion indicated that most people viewed life on the farm as less desirable than that in town. An observant foreign traveler, James Bryce, wrote in the late 1880s that "to make rural life more attractive and so check the inflow to the cities, is one of the chief tasks of American statesmanship to-day."[11]

What were the means at hand to upgrade rural living so that farming would enjoy the same attractions as other employments? The most popular suggestions were to improve rural education, to strengthen the church, to make farm life more cheerful and refined, to build better roads, and to apply mechanization and science to agriculture. These developments, it was said, would improve the image of farming and hold young men and women on the farm. Secretary of Agriculture J. Sterling Morton wrote approvingly in 1896 that farming was fast becoming one of the "learned professions" as farmers gained a better understanding of the agricultural sciences. James M. Swank, an official of the Iron and Steel Institute, declared that one no longer needed to go to the cities to find the advantages of "polite society, good schools, lectures and all home comforts and luxuries." The railroads, Swank wrote, brought city goods and ideas to the rural towns," and the wide-awake farmer and his wife and children meet city influences whenever they go to town." The clear implication was that one could remain in the country and still enjoy city advantages.

The use of machinery and improved farm practices to lighten farm labor and increase productivity made substantial gains in the late nineteenth century. These developments may have made farming more desirable, and even more profitable, but they had no noticeable effect on stopping the drain of population from the countryside. Indeed, they may have encouraged farmers to leave the fields for the factories inasmuch as increased efficiency in agriculture reduced the need for farm labor. For example, it took 39 man-hours to produce 40 bushels of corn in 1855, but by 1894 only 15 hours were required. On the average, one worker could care for only 12 acres of crops in 1850, but by the 1920s one farm laborer could handle 34 acres, or nearly three times as much. The arguments of those who believed that they knew how to keep people on the farm simply were not accepted by farmers. Even after horse-drawn machinery came into common use and scientific principles were applied to agriculture, along with the introduction of better roads, telephones, rural free delivery of mail, and other improvements, farm life could not compete with the attractions of urban living. Why?

People left the farms mainly because farming did not pay as well as nonfarm employment in a rapidly industrializing America. This is not to deny that some farmers made money. A substantial number of them prospered, both from their farming operations and from increasing land values. There were many comfortable and satisfied farmers in the United States. In 1900 some 65 percent of the farmers owned their own farms. However, overall returns to agricultural labor and capital did not compare favorably with most other economic activities in the United States. Supporters of the agrarian tradition tried to counteract or explain away the low returns from farming, but without much success. Benjamin F. Thomas of Boston declared in an address in 1862 that if a man wanted competence, "no shares—factory, bank, or railroad"—would pay better dividends than "ploughshares."[12]

People, however, could not be convinced that farming was generally a profitable enterprise. Complaints of inadequate profits fill the writings and speeches of nineteenth-century Americans. Commenting on the agrarian tradition, a New York farmer wrote in 1849 that "agriculture may be made the most happy pursuit of man" but it certainly was not the most profitable. A writer in the *Richmond* Virginia *Enquirer* estimated that "labor employed in commerce and manufactures, . . . pays three or four times as much as farming labor." In discussing the agricultural census of 1860, Director Joseph C. G. Kennedy wrote that while farm profits had generally been "steady and sure," they had not been large, and "the best talent of the country found greater attraction in other pursuits." Kennedy explained that while the United States had made tremendous economic gains in the previous generation, those advances had not favored agriculture.

Similar statements and opinions were commonplace. One observer wrote in 1872 that with lower prices and continuing high costs of operation, there had been a "large reduction" in farm profits. "This seems conclusive evidence," he continued, "that farming is not considered on a par with other branches of business at the present time." One farmer wrote in 1877 that no occupation could compare with farming "in the opportunity that it offers for *losing* of money."[13] A German immigrant in Wisconsin wrote in 1884 that he wished he could have done something besides farm. "Had I been able to work at my trade," he wrote, "I should have earned three times as much. If I had known in the beginning what I know now, I would have gone immediately to the city and kept out of the mud."

In 1890 Jacob Funck of Fairfield, Iowa, wrote a series of articles for the *Prairie Farmer* on why farm youths left their homes. The main reason, he insisted, was that towns and cities offered much better economic opportunities. Farm boys, he wrote, saw doctors, lawyers, businessmen, and others making more in a week than they earned in a month.

Statisticians confirmed what farmers and their supporters said and believed about farm profits. In 1880 income per worker in agriculture amounted to only $252 annually, compared to $572 for nonfarm workers. Moreover, the gap widened during the following decades. By 1900 the average annual income of nonfarm workers was $622 while those on the farms received only $260. That farming was not as profitable as most other types of employment was not merely the perception of farmers and their spokesmen, it was a fact. The basic unprofitability of farm work undoubtedly was the major factor in causing farm youth to migrate to the towns and cities.[14]

During the late nineteenth century, Grangers, Greenbackers, Populists, and a host of other farm groups complained about the unfavorable economic position of farmers. They charged that the growing economic and political power of industry, transportation, and finance denied farmers a fair return on their labor and investment. These farmer critics insisted that high interest rates, heavy transportation charges, marketing abuses, and political manipulation by powerful business interests were robbing farmers of their fair share of the nation's income. By the late 1800s and early 1890s, rural unrest had become intense and widespread throughout the western and southern farming regions. In 1892 farmers organized the Populist Party, and ran James B. Weaver of Iowa in an unsuccessful campaign for president on a platform calling for free silver, government ownership of the railroads and the telephone and telegraph lines, a graduated income tax, and democratic political reforms. The campaign for justice and relief for farmers reached its climax in the presidential campaign of 1896, when William Jennings Bryan of Nebraska sought the presidency as a champion of the agricultural interests.

To uphold the agrarian tradition and defend farmers against the exactions of their exploiters, Bryan engaged in a whirlwind campaign against Republican candidate William McKinley. The central issue in the election was whether silver or gold would be at the base of the national monetary system. These metals, however, were more symbol than substance. Silver represented those discontented elements in agriculture which were demanding inflation to relieve their debts, while the gold standard was symbolic of business and industrial interests. McKinley's victory in 1896 was important because for the first time in American history it became clear that the shift in economic power away from agriculture had been confirmed at the ballot box. Thus nearly a generation before farmers and rural residents became a minority of the population, they had lost out both economically and politically.

The weakness of agricultural political power continued to center around the failure of farmers to unite and organize effectively. Individualistic by nature, widely scattered physically, busy with their work, and having different economic interests, they were unable to organize and exert political power proportionate to their numbers. This was clear in the campaign of 1896, when most southern and western farmers voted for Bryan but a majority of those in the Midwest and the Northeast favored McKinley. Farmers in such important agricultural states as Illinois, Indiana, and Ohio supported the gold standard candidate because they saw no particular advantage in free silver, Milton George, editor of the *Western Rural* in Chicago, wrote in the late 1870s that if farmers would get together and organize, they could achieve political results. He declared that the legislatures were made up of lawyers and "partisan blow-hards" who were controlled by big business and who would do nothing for unorganized farmers.[15] Other farm spokesmen, as well as Grangers and Populists, saw the need for organization to achieve power, but they were unable to gain that goal in the nineteenth century and were only partially successful in the twentieth.

In surveying their economic situation, farmers did not complain that they were not as well off as their forefathers. Indeed, most of them recognized that farm life had improved significantly from generation to generation, at least throughout most of the country. What they objected to was their inability to keep up with the advances enjoyed by the rest of American society. President Theodore Roosevelt described the situation accurately in 1909 when he said: "It is true that country life has improved greatly in attractiveness, health and comfort, and that the farmer's earnings are higher than they ever were. But city life," he added, "is advancing even more rapidly."[16]

After a generation of hard times, conditions among farmers began showing improvement in the opening years of the twentieth century. Farm prices rose, land values greatly increased, and a much more favorable relationship developed between the prices farmers received for their products and those they paid for nonfarm goods. Taking the farm price index in 1899 as 100, the value of farm products rose to 133 by 1905 and to 186 in 1909. In other words, farm prices nearly doubled in about a decade. But most important to farmers was the fact that nonfarm prices advanced much less rapidly, thereby giving farm products increased purchasing power. The price relationship between farm and nonfarm prices became so advantageous to farmers that the years 1909 to 1914 came to be designated as the base period against which to judge fair or parity prices for farmers. The years between the Spanish-American War and World War I became known as the Golden Era of American agriculture.[17]

There were several reasons for the improved agricultural situation in the early twentieth century. In the first place, agricultural expansion slowed down as the best lands for farming without irrigation had been occupied. Consequently, production also leveled off. At the same time, increased demand for farm products by the nation's growing urban population and industries created a healthy market, and overseas shipments of farm commodities continued fairly strong. Domestic and foreign requirements removed excess production from the market that might have depressed prices, a common condition in the late nineteenth century.

As the best lands were occupied and demand for farm products advanced, land values rose dramatically. In some states the average price of farm land increased between 200 and 300 percent during the decade after 1900. This improved the capital position of many farmers who found that their borrowing power increased substantially.

Despite increased farm prosperity after 1900, many observers believed that agriculture was backward and out of step with the nation's growing industrialism. A group of mainly urban reformers promoted what came to be called the Country Life Movement. The main objectives of the Country Lifers were to make farming more efficient and to increase the attractiveness of farm living. While these were creditable objectives, the reformers were less interested in the farmers' welfare than they were in cheap food and in reducing the cost of living for urban residents.

The most prominent person connected with the Country Life Movement was President Theodore Roosevelt. Although Roosevelt had a strict urban background, except for his short stint on a North

Dakota ranch, he believed deeply in the main tenets of agrarianism. He favored federal help to reclaim arid western lands because this would provide more farm homes. Roosevelt repeatedly expressed the belief that farmers were special people. "The farmer represents a peculiarly high type of American citizenship," he once said. He talked about farmers being "typical" Americans and argued that the growth of cities was good but not if they grew "at the expense of the country farmer."

In August 1908, President Roosevelt established a Commission on Country Life to make a thorough study of rural conditions throughout the United States. The "problem of country life is in the truest sense a national problem," he said. Reporting in January 1909, the commission declared that "affairs in the open country" were improving, but that serious difficulties still faced the nation's farmers. The fundamental problems confronting country life, said the commission, were how to make agriculture "yield a reasonable return to those who follow it intelligently," and how to make farm life "permanently satisfying to intelligent, progressive people." There was a continued strong tendency, the commission said, for farmers to move to towns and cities, indicating that country life did not satisfy many rural residents. The most serious problem, reported the commission, was that farmers stood "practically alone against organized interests." The "inequalities" and "discriminations" that industry, finance, and transportation companies had forced on farmers resulted from "lack of a highly organized rural society." If farmers were to protect their interests, organization for self-interest was essential.

The commission recommended many of the same old panaceas that had been discussed for a generation or more—better schools, improved roads, parcel post, more agricultural credit, diversified farming, establishment of cooperatives, and government control of monopolies.[18]

The Country Life Movement failed to catch the attention of most Americans. Even farmers themselves did not show much interest in it. Yet within a few years a number of the movement's objectives had been achieved. Parcel post became a reality in 1910. The Clayton Antitrust Act of 1914 encouraged the formation of farmer cooperatives by providing some protection against prosecution. Enacted in 1916, the Federal Farm Loan Act expanded agricultural credit. In 1916 Congress also passed the Rural Post Roads Act, which provided federal matching funds for so-called rural post roads. Of this law President Woodrow Wilson said that it would "promote a fuller and more attractive rural life." Agricultural education through extension and short courses received federal support in the Smith-Lever Act of 1914, and in 1917 Congress enacted the Smith-Hughes Act, which distributed federal funds to encourage instruction in vocational agriculture and home economics in the nation's secondary schools. Appropriations for the growing functions of the United States Department of Agriculture rose sharply in the early years of the twentieth century. In 1899 Congress appropriated only $2.8 million for the USDA; by 1917 the figure reached $28 million. President Wilson believed that Congress's support for agriculture had strengthened the country's "great agricultural foundations."

No amount of individual, group, or federal effort, however, could change the inevitable tide. Relatively, rural America was on an irreversible decline. Within little more than a century the farm majority had become a minority. The census of 1920 showed that only 29.9 percent of Americans still lived on farms. The South and Midwest had the highest percentage of people on farms, but even those regions were experiencing a decreasing farm population. Mississippi, Arkansas, and North Dakota were the most rural states. They had 71, 63, and 61 percent respectively of their people on farms in 1920. Despite a declining farm population, the number of farms remained about the same, around 6.5 million, between 1915 and 1935. Since these millions of farms varied greatly in size, efficiency, profitability, and production patterns, it is now time to examine the nature and character of American farming as it existed in the early twentieth century.

American Farms and Farmers

II

BY THE EARLY TWENTIETH CENTURY AMERICAN farmers had developed the most productive and varied system of agriculture that the world had ever seen. A great variety of grain, cotton, vegetable, and fruit crops, as well as livestock, poured from the country's farms and ranches to supply the needs for food and fiber both at home and abroad. There was no "typical" American farm or farmer. Some operations were general, producing several crops and different kinds of livestock and poultry. Other farmers specialized, concentrating on one or two crops or raising a particular kind of livestock. Many farms were small, covering less than 50 acres, while others extended over hundreds and even thousands of acres. For the country as a whole the average in 1920 was 148 acres. There were small, commercially unproductive farms, such as those in the Appalachian region, that were largely self-sufficient, while the large grain and livestock enterprises in the Midwest were examples of a high degree of efficiency and commercialism. The variation in American farming was almost endless.

The people occupying the 6.5 million farms in 1920 were mainly white. There were, however, some 925,000 black farmers, concentrated mainly in the lower South. The legacies of slavery, racism, and poverty rested heavily on this group. Most of them labored on small farms as tenants or sharecroppers and lived a marginal existence. In some states the average acreage worked by black farmers was only about half as much as that of their white neighbors. Thousands of black farmers could not be considered viable commercial operators at all. They were an impoverished minority within a minority.

Over time American agriculture had settled into regional patterns determined by such factors as soil, climate, and markets. In New England the thin rocky soil, the short growing season, and competition from western producers had discouraged staple-crop farming as early as the eighteenth century. In the nineteenth century the declining number of New England farmers turned to producing specialized commodities for the growing urban markets. Dairying became the leading agricultural activity in New England, and by the late 1930s these enterprises earned about 34 percent of the region's gross farm income. Poultry and poultry products ranked second in importance. Maine potatoes became famous nationwide, and New England farmers produced a wide variety of other vegetables, as well as fruits and nuts. Massachusetts cranberries were a major specialty crop. Thus the principal activities of New England farmers centered around hay and pasture for dairy cows, and poultry, vegetable, and fruit production. But, overall, farming in New England was a relatively small part of the region's total economic activity, and the farm population declined to only 8.5 percent of the total by 1920 and continued downward in later years. The average size of farms in New England was only 98 acres in 1940.

The Middle Atlantic states of New York, Pennsylvania, and New Jersey made up another agricultural region where nearby urban markets helped to determine the major farming activities. Historically, the area's farmers produced wheat and other grain crops for the domestic and foreign markets. But even with better soils and somewhat larger operations than found in New England, farmers in the Middle Atlantic states were unable to compete with western staples by the late nineteenth century. They, too, turned to products that could be marketed profitably in such growing cities as New York, Buffalo, Philadelphia, and Pittsburgh. The most important farm enterprise was dairying. While farms were small, averaging slightly under 100 acres in 1940, many of the operations were very intensive and the output per acre was high. Indeed, New York was the nation's leading dairy state until surpassed by Wisconsin about 1912. By 1939 the value of New York's dairy products exceeded $100 million and dairying was the most important source of farm income. Livestock, including dairying, accounted for some two-thirds of total farm income in the Middle Atlantic states by the late 1930s. Hay was the principal field crop.

The remainder of the region's agriculture consisted of vegetables, poultry, and horticultural products. Because of its soils and proximity to markets, New Jersey became one of the country's leading producers of vegetables both for the fresh markets and for canning. New York's output of potatoes and apples was among the highest in the country. Specialty crops such as tobacco and mushrooms grown in particular localities also added to farm income. While many crops and different kinds of livestock were grown in New York, Pennsylvania, and New Jersey, dairy, poultry, and vegetable farming predominated. Farmers had found what they could produce and market most efficiently, and concentrated on those activities. As in New England, the number of people living on farms in the Middle Atlantic states declined rapidly in the nineteenth and early twentieth centuries. By 1940 only 6.5 percent of the region's people lived on farms.

Another agricultural region, or more accurately a subregion, that developed distinctive qualities was the area from Delaware to the Virginia tidewater. Known locally as Delmarva, it was a small region of highly intensive farming with emphasis on dairying, poultry, and vegetables. By 1939 little Delaware marketed more chickens than any other state in the Union. Maryland was also a heavy producer of poultry. However, in Maryland dairying became the leading agricultural activity. Much of the Delmarva region was in the Atlantic Coast truck-farming belt, which extended from New England to South Carolina. Such crops as tomatoes, sweet corn, cucumbers, English peas, and other vegetables occupied large acreages. In tidewater Virginia peanuts grew well and the small community of Suffolk boasted of being the "Peanut Capital of the World." Overall, the region concentrated on dairying and poultry and raised a large variety of vegetables and fruits. The farming patterns were not much different than in parts of the Middle Atlantic states and New England, where nearby urban markets influenced agricultural development. Northeastern farmers, however, were not without problems. Production of milk, poultry, and vegetables sometimes exceeded market demands and depressed prices to unprofitable levels.

One of the poorest farming regions in the United States developed in the Middle South. Consisting of western Virginia, North Carolina, West Virginia, Kentucky, and Tennessee, this region had 1,025,000 farms by 1930, nearly one-sixth of the nation's total. Tucked away in the mountains and valleys that make up so much of the area, these enterprises were mostly small, and unproductive. The average farm was around 80 acres. Approximately one-third of the farms in this region were under 30 acres, and thousands of farmers in the 1920s and 1930s harvested less than 10 acres of cropland. On many of these acreages farmers produced barely enough for their own use and had little or nothing to sell in the commercial markets. The Appalachian area of the Middle South, consisting of eastern Tennessee and Kentucky, western Virginia, much of West Virginia, and the extreme western part of North Carolina, was probably the most self-sufficient agricultural region and also one of the poorest. Corn was the principal crop. It was eaten in many forms and also distilled into whiskey. Farmers also raised small grains, hay, and livestock, but on a very limited scale.

In some parts of the region, mainly outside of Appalachia, commercial production of major crops prevailed. In southwest Tennessee and in parts of North Carolina cotton was the major money crop. Tobacco was produced in Virginia, eastern North Carolina, and in Kentucky and Tennessee. In North Carolina tobacco exceeded all other crops as an income producer. In the Shenandoah Valley of Virginia grains and livestock were raised on general farms, and that area also developed commercial fruit production, with emphasis upon apple growing. While the region raised a variety of crops and livestock, and had a few prosperous farmers, agricultural developments remained fairly static and relatively uninfluenced by modern farm practices. For example, only about 4.5 percent of the farmers in the Middle South had tractors in 1940 compared to over 53 percent in the North Central states. The small, irregularly shaped, hilly farms simply did not lend themselves to modernized farming. Moreover, the region's rural poverty did not permit the accumulation of capital necessary for improved agricultural techniques. Much of the land was seriously eroded, and the effect of the Tennessee Valley Authority on soil conservation was not felt to any substantial degree before World War II. The basic problem in the region was the pressure of people upon the land. Land resources were simply not sufficient to produce a decent living for so many people.

The Deep South, often referred to as the Cotton Belt, stretched from South Carolina to central Texas. For a thousand miles or more, through South Carolina, Georgia, Alabama, Mississippi, Arkansas, Louisiana, and Texas, cotton was the leading money crop. Of the 2.3 million farmers in the region in 1920, some 74 percent of them raised cotton. The white lint dominated southern farm life in the early twentieth century just as it had in the nineteenth century. In 1930 the southern farms that were classified as cotton farms by the Bureau of the Census ranged from a high of 82 percent in Mississippi to a low of less than 10 percent in Florida. By the 1920s southern farmers were producing anywhere from 11 to 13 million bales annually. When cotton prices were good, there was great joy throughout the South, but when prices dropped, or the boll weevil destroyed part of the crop, grinding poverty and despair prevailed.

The sharecropping, crop lien, and other systems of farm tenancy that developed in the South after the Civil War contributed greatly to fastening cotton culture on much of the region. While the soils and climate of the South were well suited to the crop and the cash return was probably higher per acre than for most other field crops, emphasis on cotton production resulted to a large degree from the fact that landowners and creditors insisted that it be grown. The advantages of cotton from the landowner's or creditor's point of view were that it always had a cash market and it could not be pilfered or eaten by the farmer.

While cotton was the premier crop in the Deep South, corn occupied more acres of farmland. Corn, however, was not a cash crop. It was eaten by people and livestock. Such staple corn products as grits, hominy, mush, and cornbread made up much of the southern diet. Farmers also raised other grains and livestock. In Arkansas, Louisiana, and Texas rice became an important crop, while Louisiana planters grew a good deal of sugarcane. Fruits and nuts were also widely grown. Georgia and Alabama peaches became famous nationwide, as did peanuts and pecans, and Florida citrus had become important in that state. Southern farmers also raised a wide variety of vegetables. But the emphasis continued on cotton.

During the late nineteenth and early twentieth centuries, agricultural reformers criticized the concentration on cotton to the exclusion of other crops. Farmers were advised to diversify their crops and raise more and better quality livestock. This meant substituting grains, grass, and vegetables for cotton, and raising hogs, chickens, and sheep. The agricultural reformers argued that it was economic stupidity to buy basic food products when farmers could raise them. Why buy bacon or butter when they could be produced on the farm? George Washington Carver, the black agricultural scientist, wrote in 1902 that "it is not unusual to see so-called farmers drive to town weekly with their wagons empty and return with them full of various kinds of produce that should have been raised on the farm." Most advice to diversify fell on deaf ears, and the great majority of farmers continued in their old ways through the 1930s.

The economic condition among most cotton farmers was little above the subsistence level. In the late nineteenth century annual incomes of from $100 to $150 were common among both black and white sharecroppers. For many it was less. Conditions improved a little in the early twentieth century, but even in the so-called prosperous 1920s many poor farmers in the South existed on less than $200 cash a year. In 1938 President Franklin D. Roosevelt referred to the South as the nation's number one economic problem. Much of that problem was on southern farms. While farmers in the region faced many problems, the main difficulty was their low productivity per man-hour. In short, the South had a substantial surplus of labor that could not be profitably employed in agriculture. There were too many farmers in relation to land resources. Southern farmers also lacked capital.

While every part of the United States could boast of rich agricultural resources, no region of the country could compare with the Midwest, that vast region stretching more than a thousand miles from central Ohio to eastern Nebraska. It was no exaggeration to call this great heartland of America the richest farm on earth. Because of the predominance of corn cultivation and hog feeding, the region became known as the Corn-Hog Belt. But good soil, temperate climate, and adequate rainfall combined to encourage the production of a great variety of crops, livestock, and poultry. In the 1920s and 1930s farmers began to add soybeans to their main field crops.

Besides corn, which was raised on about 80 percent of Midwest farms in 1939, farmers produced oats, wheat, barley, rye, and soybeans. Except for wheat, most of the grain was fed to millions of hogs and cattle. Ranchers from the states farther west shipped thousands of calves into the region annually to fill the feedlots of Iowa and nearby states. On the northern edge of the Corn-Hog Belt, dairying became the leading farm activity. Wisconsin was the premier dairy state, but Minnesota and Michigan were also large milk producers. In those states, except for southern Minnesota, which was a part of the corn empire, grass and hay were leading crops. Vegetables and fruits also did well in the Midwest. Farmers raised large quantities of such vegetables as tomatoes, sweet corn, and potatoes, as well as apples and peaches in Ohio and Illinois, grapes along the lower Missouri and Ohio rivers, and berries in Michigan. It was a varied and productive agricultural economy. The region's uniqueness, however, was not in its variety of crops and animals, but in the special relationship between abundant grain production, especially corn, and livestock. By converting grain into meat and dairy products, midwestern farmers provided most Americans a tasty, high-protein diet.

The generally level terrain of the region and the size of farms invited mechanization. By 1930 midwestern farms averaged about 130 acres each. While farmers in the Midwest had used horse-drawn machinery on an ever-increasing scale since the late nineteenth century, during the 1920s they turned rapidly to tractor power. Some 30 percent of the farmers in Iowa and Illinois had tractors by the end of the decade, nearly double the figure in 1920. Between 1894 and 1930 the man-hours required to produce an acre of corn declined from 15.1 to 6.9. Mechanical advances, improved crop strains, and selective breeding of livestock all added to the richness and productivity of midwestern farmers, who numbered 1,622,625 in 1930.

Farther west farmers had developed a kingdom of wheat and cattle. The western prairies and Great Plains extended over a vast area from Texas to the Canadian border. Much of this semiarid area suffered from periodic drought, which required successful farmers to summer-fallow, to grow drought-resistant crops, and to balance their operations between grain and livestock. Even the best farm practices, however, could not protect farmers from periodic disaster in the Wheat Belt. From the early settlements of the 1870s to the Dust Bowl days of the 1930s, some farmers concluded that success was impossible and abandoned the region. Despite the heartaches and bankruptcies, farmers on the Great Plains who stuck it out eventually adjusted to the erratic climate and made the region one of the world's great breadbaskets.

While general farming characterized the eastern parts of Kansas, Nebraska, and the Dakotas, wheat became the dominant crop throughout the Great Plains. Kansas and North Dakota were the nation's leading wheat states, usually producing from one-fourth to one-third of the 800-million-bushel crop that was commonly raised in the 1920s. Farmers in those two states devoted half or more of their harvested cropland to wheat. The suitability of the soil, the topography of the land, and the early development of machines to handle all aspects of wheat farming contributed to some huge wheat operations involving hundreds and even thousands of acres. The average farm in the Dakotas in 1930 was between 400 and 500 acres, while the figure reached 345 acres in Nebraska and 282 acres in Kansas. South Dakota's average farm was nearly five times larger than farms in the cotton state of South Carolina. Production efficiency grew rapidly on most wheat farms. The man-hours required to produce an acre of wheat fell from 8.8 to 3.3 between 1894 and 1930. Plains farmers were in the vanguard of those who early forsook horses and turned to tractor power. By 1930 between 30 and 40 percent of Great Plains wheat farmers used tractors, and in North Dakota the figure was nearly 44 percent.

While wheat was the single most important grain raised on the western prairies and Great Plains, it was not the main crop throughout all of the region. In southeastern South Dakota, eastern Nebraska, and Kansas, corn was the principal grain. These farmers were really part of the Corn Belt. They raised hogs and fattened cattle just as farmers did who lived farther east, in Iowa and Illinois. Other grains raised included oats, rye, barley, and milo. Hay and flax were also important crops. Many farmers had dairy cows and chickens, but these were mostly side enterprises which produced some additional income from cream and egg sales.

The cattle industry was a major agricultural activity for thousands of the region's farmers and ranchers. The Great Plains could boast of being one of the world's greatest grazing regions. Indeed, cattlemen in the 1870s and 1880s were among the first settlers to exploit the Great Plains' most valuable natural resource—grass. After open-range ranching declined in the 1890s, ranchers fenced their lands and raised tens of thousands of cattle behind miles of barbed wire. The fine grasses of the Flint Hills of east-central Kansas, the Osage ranching country of northeastern Oklahoma, and the Sand Hills in northwest Nebraska made those areas especially famous for ranching. Texas, of course, was the premier cattle state, a reputation it had earned even before the Civil War. In 1930 Texas farmers and ranchers reported 6.6 million head of cattle, more than 10 percent of the nation's total. Income of the more than one million farmers on the western prairies and Great Plains came mainly from livestock and wheat.

Farming and ranching farther west, in the Rocky Mountain states, had been determined by terrain and climate. Much of New Mexico, Colorado, Wyoming, Idaho, and Montana was arid and required irrigation for successful crop production. Melting snows provided water for crops in the rich river valleys just east of the mountains and in the intermountain areas. There farmers developed a highly intensive agriculture. In Colorado and Utah sugar beets became a major cash crop, and in Idaho and Colorado farmers raised large quantities of potatoes. Farther south, along the Rio Grande, such specialties as pinto beans and chili peppers became important. Fruits and vegetables were other cash crops, and thousands of acres were devoted to alfalfa and hay for the thriving livestock industry. However, only a small portion of the land area was in harvested cropland. The Rocky Mountain region was the realm of cattlemen and sheepmen. Lambs and calves were the main products streaming from the area's farms and ranches. Much of the land remained in the public domain, and the federal government leased pasturage to the ranchers. A common pattern was for ranchers to own valley land, where they grew winter hay and feed while letting their livestock graze over the leased plateau and mountain ranges in the summer.

Because of emphasis upon livestock, farms and ranches were relatively large in the Rocky Mountain states. By 1930 farms in New Mexico, Colorado, Wyoming, Idaho, and Montana averaged slightly above 800 acres. Some of the irrigated farms were small, but the number of large ranches boosted the average farm size to the highest of any area in the United States. The five states had a total of less than 200,000 farms in 1930. These farmers and ranchers depended mainly on livestock for their income.

Between the Great Divide and the Pacific Ocean, a tremendous variation developed in the size of farms, the kinds of crops, and in farm organization patterns. For much of the region, with its arid expanses, water was the key to agricultural development. Without irrigation most of the western slope would have remained in the hands of ranchers and a few dry-land farmers. However, by the early twentieth century, and in some cases much earlier, extensive irrigation projects were developed in Arizona, California, Utah, Oregon, and Washington. These water facilities opened the way for transforming hundreds of thousands of additional acres of land into productive farms to grow a wide variety of staple and specialty crops. For example, the Salt River Valley project in Arizona, completed in 1911, permitted much of the desert in that state to grow crops.

The variation in far western agriculture was almost endless. In western Washington and Oregon, where rainfall was heavy, dairying developed to supply the milk needs of growing cities such as Portland and Seattle. In the Willamette Valley of central Oregon rainfall permitted farmers to grow grains, hay, fruits, and vegetables. In the more arid sections of eastern Oregon and Washington large wheat farms and cattle ranches prevailed. The Palouse area of eastern Washington and western Idaho was one of the country's greatest wheat-producing regions. Farms were large and highly mechanized. Washington, as well recognized for apples as Mississippi was for cotton, was also a major producer of pears, cherries, and other fruits. Nevada was mainly a thinly populated ranching state.

California, however, was the West's richest agricultural empire. In 1929 it ranked second among the forty-eight states in the value of farm products sold, exceeded only by Texas. The main field crops were wheat, barley, rice, and cotton. Although wheat acreage dropped in the years after World War I, California farmers sharply increased their plantings of cotton. This was possible because of the expanding irrigation facilities which were necessary to grow cotton in the arid valleys. It was the production of citrus fruits—oranges, lemons, grapefruit, and limes—that attracted national attention to the state. Between 1919 and 1929, California fruit growers increased their production of oranges from 21.6 million to 43.1 million boxes. California farmers also grew grapes, a great variety of nuts, vegetables that were being shipped fresh to eastern markets, and other specialties. By the 1920s the Golden State was supplying a substantial part of the nation's markets with fresh fruit and vegetables. In 1929, the state grew 75 percent of the nation's oranges and 15 percent of the grapefruit. Dairying and beef cattle also contributed to farm income in California. At the same time, irrigation waters were becoming more available in Arizona, where farmers were increasing their production of vegetables, citrus, and, a little later, cotton.

As one viewed American farming from New England to California and from North Dakota to Texas in the years after World War I, the picture that emerged was that of abundance and variety. Farmers raised about every staple and specialty crop found outside of a tropical environment, and they had developed the greatest livestock economy on earth. American farms supplied more than enough to feed the country's growing population, although not everyone had adequate income to buy sufficient food. Moreover, agricultural products continued to furnish the raw materials for a number of leading manufacturing enterprises—cotton and wool for the textile mills, livestock for the meat-packing plants, grain for the flour industry, tobacco for cigarette manufacturers, and many others. Farm products were also highly important in America's world trade. In 1929 nearly $2 billion worth of agricultural commodities were exported. This was 35 percent of the nation's total exports. Overall, American farms were responsible for about 10 percent of the national income.[1]

When considered from the viewpoint of aggregate production, American farmers had been highly successful. Indeed, the heart of what came to be known as the "farm problem" was excessive production and surpluses, which depressed prices below profitable levels. However much the country as a whole may have benefitted from a productive farming industry, the economic rewards going to farmers were low. In 1929, a supposedly prosperous year, the net income of the farm population averaged a meager $273 per capita compared to the national average of $750. The average full-time farm worker earned only $378 in 1929 as against the average for all industries of $1,405. The United States Department of Agriculture surveyed 11,851 farm families throughout the nation in 1928 and found that only 20 percent of them had incomes of $2,000 or more. Twenty-five percent fell into the $1,000–$2,000 bracket, and 45 percent of the farm families studied had annual net returns from farming operations of less than $1,000. Twenty-three percent made less than $500. These meager returns were in many cases for labor of entire farm families. By any calculation, income to farmers on their labor and capital was small. As Louis H. Bean, senior agricultural economist in the USDA, wrote in 1930, "farmers have been moving to industrial centers because farm earnings have not been satisfactory."[2]

Economic and living conditions varied greatly among farmers.

The poorest were in the South, where hundreds of thousands eked out a bare living. Operating as tenants and sharecroppers, heavily in debt, and with low production, both white and black farmers experienced extreme poverty. They lived in leaky shacks, existed on a diet mainly of cornmeal and fat pork, suffered from malnutrition, pellagra, and hookworms, and lacked education and other advantages in life. Only occasionally were cotton prices good enough to provide a temporary respite from their hard lives. On the other hand, many farmers in the Midwest and elsewhere enjoyed a fairly comfortable existence. They cultivated larger and more productive farms, diversified their crops and livestock, managed well, and made enough to meet the necessities of life, and more. Their homes were large, well painted, and comfortable, they had good outbuildings for livestock and grain storage, and they earned enough money for occasional travel and to provide some education for their children. They had books and periodicals, telephones, a piano or an organ, and other amenities of life. Some of the larger farmers and ranchers could claim real prosperity. But they were the exception rather than the rule.

Several factors accounted for the differences in income and living standards among the nation's farmers. The size of farm, the quality of the soil, the availability of capital, the amount of debt, the kind of crops and livestock raised, and the level of education and management abilities were all important. Some conditions were within the power of individual farmers to modify or change. Others were not. It was nearly impossible for a poor Mississippi sharecropper who farmed 20 or 30 acres to improve his condition. He had no access to the necessary capital to buy land or to change his farming operations from cotton, which in many years was in oversupply, to some other crop or livestock. Moreover, he lacked education and managerial skills. Many midwestern farmers, on the other hand, could improve their efficiency through mechanization, or a shift in crop and livestock production, and in general could change with the economic tides.

There were some basic matters, however, over which farmers had no control whatever, either individually or as a group. One of these was the general level of business and industrial growth, which influenced the demand for farm products. During periods of sluggish business activity, farm prices not only fell but usually fell more than other prices. One of the most important reasons for the sharp decline in farm prices after 1920, with a drop of nearly $6 billion in gross farm income between 1920 and 1921, was the nationwide business recession. The worldwide demand for American farm products was also vitally important to farmers, especially to cotton producers and wheat growers, who exported a substantial part of their crop. Some 60 percent of the 1927–28 cotton crop and 21 percent of the 1927 wheat crop were shipped abroad. Low world prices for these products could cause great distress on hundreds of thousands of American farms.

But the fundamental problem facing farmers was the unfavorable relationship between farm and nonfarm prices. What mattered to a farmer was the exchange value of his products. If he could exchange his products through the price mechanism for a fair supply of nonfarm commodities that he must buy for his business operations and living, he would be reasonably well off providing he had products to sell. However, if his prices were low and nonfarm prices were high, he experienced a sharp disparity in the exchange. "The failure of the prices of farm products to decline and to rise at the same rate as the prices of other products," one agricultural economist explained in 1924, "has turned out to be one of the most important maladjustments of the business cycle."[3] What farmers wanted was parity between the prices of their products and the commodities they had to purchase in the nonfarm sector.

After considerable study of price relationships, the United States Department of Agriculture had determined that in the period 1909 to 1914 prices of farm products were at parity with the prices of nonfarm commodities. That is, during those years a certain farm commodity could be exchanged in the marketplace for a fair amount of nonfarm products. This fair relationship was later dubbed 100 percent of parity. Subsequently, farmers and their spokesmen looked back to 1909–14 as the benchmark against which to compare and assess their economic condition. Whenever farm prices fell disproportionately to the prices of other commodities, farmers received less than 100 percent of parity and lower prices than they believed they deserved.

Why did farmers find themselves in a condition of economic disparity? The main reason for their unfavorable economic position was that they had little or no control over the price they received for their products or what they paid for nonfarm goods. When they took their crops and livestock to market, they had to take a price set by others. That price was determined by general market conditions and had no relation to the need of farmers for a price that would cover production costs and profits. Indeed, those who established farm prices were concerned with profits for the purchaser, not the producer. When the farmer took his wheat to the elevator or his cotton to the gin, he had two choices—take the price offered, or return home with his product. Individual farmers were in no position to bargain effectively with grain elevators, meat processors, or other purchasers of farm commodities. It was a take-it-or-leave-it situation for the farmer.

When farmers went to purchase machinery, fertilizer, and other products that entered into their cost of operations, as well as items for family living, they had to pay the price asked by the retailer. Again, farmers were unable to bargain for more favorable terms. Assume, for example, that the price of wheat declined under the pressure of a large crop or declining exports. When the wheat farmer went to buy farm machinery, he might explain to the dealer that because wheat prices were down farm implement prices should be reduced accordingly. But it did not work that way. While those selling commodities used by farmers extended occasional discounts, they generally maintained their prices at profitable levels regardless of the trends in farm prices. The farmer had no economic clout with the machinery dealer or others selling things to him. Thus farmers usually paid a price that included a profit for the manufacturer, wholesaler, and retailer, regardless of any effect this might have on the farmer's profit or standard of living. In short, farmers bought at retail and sold at wholesale. As many farmers expressed the situation, they sold low and bought high.

What was responsible for this condition among commercial farmers? The main reason was that farmers did not control their production and therefore the prices of their products. Farmers did not get together and try to determine the amount of a particular agricultural commodity that would assure them a fair price on the basis of supply and demand. Indeed, farmers paid little attention to what other farmers were doing. If American farmers had any outstanding characteristic, it was their strong individualism. But even if farmers had attempted to gauge their production to effective and profitable demand, their production goals would have been altered by insects, droughts, floods, and other natural disasters. They simply could not control output with the precision of a manufacturer. Secretary of Agriculture Henry C. Wallace explained this in his annual report of 1922 when he wrote: "It will never be possible for the farmers to relate their production to profitable demand with the nicety of the manufacturer, both because they can not control the elements which influence production and can not estimate demand as closely."[4] Most farmers produced blindly, paying little attention to aggregate demand and the influence that quantity had on prices.

How could farmers change their basic position in the economy and improve their bargaining power in the marketplace? How could they get better prices for their own products and pay less for the nonfarm goods they had to buy? There was no single or simple answer to these questions, although a variety of solutions had been advanced throughout American history. On the production side, some farmers had periodically advocated reducing output to the point of effective or profitable demand. This idea was older than the Republic. On several occasions in the seventeenth century, Virginia and Maryland tobacco growers had urged the destruction of surplus tobacco in order to raise prices. The British Crown would not permit such a policy, but the idea remained alive. In 1845 a group of cotton planters met in Montgomery, Alabama, to consider the question of limiting cotton output until prices recovered to a profitable level. But nothing came of this and other meetings.

Another approach was for farmers to organize and deal with their problems through cooperative action. If other groups, particularly business and labor, were benefitting from organization, why should farmers continue to act alone. The Grange, formed in 1867, became a leading advocate of organizing farmers into marketing and purchasing cooperatives as a means of solving their economic problems. Two decades later the Farmers Alliance also pushed the cooperative principle as the best way to improve the condition of millions of American agricultural producers. In an effort to pay lower prices for the things they bought, farmers banded together in their cooperatives and purchased goods in quantity directly from jobbers and manufacturers, thus bypassing retailers and commission agents. Grangers and, later, Alliancemen, formed cooperatives to buy lumber, twine, machinery, and other supplies. They also established producer cooperatives to market grain, livestock, and fruit, hoping to get higher prices by cutting out the charges of middlemen. While hundreds of cooperatives were organized in the late nineteenth century, most of them failed. Lack of capital, opposition by local merchants, poor management, and the absence of a genuine cooperative spirit among individualistic farmers condemned most cooperatives to failure. Although the farmer cooperative movement made some gains in the early twentieth century, its overall influence on rural welfare was slight.[5]

Of course, farmers had recognized for many years that government action could be a very important factor in their welfare. Policies relating to land, money, and the tariff were among those that could affect the agricultural industry either for good or for ill. While passage of the Homestead Act, establishment of the United States Department of Agriculture, and enactment of the Federal Farm Loan Act were among measures helpful to farmers, most agricultural legislation had no effect on getting farmers higher prices for their products. Indeed, the United States Department of Agriculture and the federal-state experiment stations concentrated on showing farmers how to increase productivity. This contributed even further to the surplus problem. In the early 1890s, the Farmers Alliance urged the federal government to take more positive action on behalf of farmers. Alliance leaders advocated the establishment of a subtreasury system whereby farmers could borrow money directly from the national government on stored commodities. From the producers' viewpoint, this plan had the advantage of making it possible for farmers to keep their crops off a depressed market, and issuing paper money to make the loans would, it was thought, inflate the currency and raise prices. This scheme, however, never got a serious hearing in Congress. As farmers continued under economic pressure, they complained about government favors to business and industry, and its neglect of agricultural interests.

The problems of farmers came into dramatic focus in the autumn of 1920 when the nation entered a sharp postwar depression. Some farm prices plummeted to less than half their previous level within a few months. Cotton brought 37 cents a pound in July 1920, but it sold for 13 and 14 cents early in 1921; wheat, hogs, and cattle also experienced disastrous declines. Large crops, a drop in exports, competition from other countries such as Australia and Canada, and general price decreases all contributed to hard times on the farm. The situation was particularly difficult because many farmers had extended their debts during World War I and had to pay interest and principal on loans with dollars coming from low-priced products. While nonfarm prices also declined in 1920 and 1921, they dropped much less than did agricultural values. As Clarence A. Wiley, a contemporary economist, explained: "agricultural prices fell first, fell fastest, and fell farthest." In terms of purchasing power, farm commodities dropped to one of their lowest levels in the twentieth century. The average for several basic agricultural commodities in 1921 was only 67 percent of parity. Secretary of Agriculture Henry C. Wallace pointed out that a suit of clothes that cost a North Dakota farmer 21 bushels of wheat in 1913 cost him 31 bushels a decade later. A wagon that could have been purchased for 103 bushels of wheat in 1913 required 166 bushels in 1921. Farmers were suffering from a harsh cost-price squeeze.[6]

Widespread hardship occurred throughout rural America in 1921 and 1922. J. J. Brown, the Commissioner of Agriculture in Georgia, wrote in July 1921 that never during his thirty-five years of close contact with farmers had he seen them "so depressed and in such an alarming financial condition as they are today."[7] Thousands of farmers had to tighten their belts and lower their standards of living, a standard that was already well below that of the nonfarm population. But even the greatest sacrifices were not enough to permit them to keep up taxes, interest, and loan payments. Many farmers suffered the final blow when they lost their farms to creditors. They then either left the farm or were added to the tenant class. Between 1920 and 1923, 8.5 percent of the owner-operators in fifteen corn and wheat states lost their farms through foreclosure or voluntary relinquishment of property to creditors.[8] An additional 14.5 percent held on only because of the leniency of those holding their mortgages. The postponement of foreclosures in the early 1920s accounted for the rise in tenancy in the middle and late 1920s, rather than in the immediate postwar years.[9]

Statistics on prices and foreclosures, however, cannot tell the story of human misery faced by so many farmers. Many of them, especially in the South, had never been prosperous, and the postwar economic slump aggravated an already bad situation. A Georgia farmer wrote to the United States Department of Agriculture that he had eight "head of children and my wife," but even with hard work he could not feed or school them. His children, he said, would be better off in an orphans' home. "My family," he wrote, is "in bad shape they are naked and barefooted." A southern housewife wrote to President Warren G. Harding in June 1921: "This fall not only will I lose my home and everything in it, but hundreds, perhaps thousands, will be in my condition, homeless."

Poor southern farmers were not the only ones who felt the postwar economic pressures. Established operators and landowners in some of the richest agricultural regions of the country faced bankruptcy. A Colorado farmer wrote in March 1921: "I have farmed or hired farming done west of the Missouri for fifty-five years, gone through drouth seasons and grasshopper seasons, but never seen as discouraging time for the farmer as at present." A few months later a Washington State farmer disclosed to Secretary Wallace that he had not been able to buy a new suit for eight years. A Texas cotton producer explained that the cost of raising cotton on his 90-acre farm was 8 to 12 cents a pound more than he received for it. "Where am I? No doubt you will think broke—which is right." Except for the high wartime prices in 1918–19, he said, producing at below cost of production "is as perpetual as the sun. . . . There are thousands and thousands of farmers driven from the farm this year because of their financial inability to stay." A Missouri cattle feeder reported that he had purchased two hundred head of feeder cattle in the fall of 1919 at what he considered a price that would produce a profit. He wintered and pastured them until September 1921, when he put them in the feedlot. He sold a hundred head in February 1921, and as fat cattle they brought $2000 less than they had cost as feeder calves. Nor did he see any prospect of getting his costs out of the other hundred. This farmer told Secretary Wallace that he had borrowed money to pay his interest and taxes and had been forced to negotiate a second mortgage on his farm.

Cases of financial desperation were evident even in Iowa, one of the nation's richest farm states. A farm wife from near Nevada, Iowa, wrote to Secretary Wallace in June 1921, explaining the family's critical situation. "We have a beautiful farm of 240 acres" which was "all tiled and fine land, black soil . . . and the improvements are good," including a newly painted house and excellent outbuildings. She explained that the farm indebtedness amounted to some $35,000, including loans from the Equitable Life Insurance Company and two local banks at interest rates varying from 5.5 to 8 percent. This farmer had tried to sell the farm to his creditors at a price that would get his investment back, or at least keep him from losing everything, but both bankers who held mortgages on the farm refused to purchase the land even at depressed prices. The farmer had sold off most of his livestock to pay interest. Now, this Iowa farm wife wrote, if they could not get a government loan they would lose everything.[10]

Secretary Wallace deplored the depression among farmers, and expressed some alarm at the possible national consequences. He wrote in 1923 that the "drift from the farms to the cities is due in part to inability to make a decent living on the farm." The situation was "draining from the country such a large percentage of the more intelligent and ambitious young farmers." The Department of Agriculture estimated that "the net change in population from the farm to the town in 1922 was around 1,200,000 persons."[11] This urban draft upon the "best the country produces," the Secretary continued, was "altogether heavier than is good for either the country or the nation." Many other Americans expressed deep concern over the economic pressures that were forcing farmers out of agriculture.

It was clear that conditions on American farms needed improvement. The main question was, what should be done? Should the farmers themselves try to solve their problems individually, get together in organized groups, or turn to government for assistance? Despite a flood of proposals on how best to solve the farmers' price problems, there was little agreement on any course of action among farmers or their spokesmen. The suggestions included diversification, establishment of more cooperatives, raising the tariff on agricultural commodities, more liberalized federal credit, the setting of prices by Congress at a level to guarantee farmers a profit, and plans to have the government buy up surplus crops and sell them abroad to keep price-depressing surpluses from ruining the domestic market.

As the Secretary of Agriculture, the President, and members of Congress were bombarded with ideas for solving the "farm problem," the central question that emerged was whether the federal government would play any direct role in raising farm prices. If this idea won support, did farmers and their spokesmen have the political power to gain legislation only of interest to farmers and which might create higher food and raw material prices for other elements in the economy? Historically, farmers had not been very successful, despite their numbers, in winning special interest legislation that might affect the price of their products. Numerically, by the early 1920s farmers had less influence than ever before in American history. They had become a distinct minority of the population. To make the problem of organizing political power even more difficult, the economic interests of farmers were not identical. In other words, the farm minority could not or would not work in political harmony. The outlook, then, did not appear particularly good for farmers to achieve any kind of price support legislation in the early 1920s.

The great majority of Americans still believed that there was something good about farmers and farming. They accepted the notion that farmers were honest, dependable, virtuous, politically stable, and wedded to liberty. Furthermore, many citizens, not just farmers, believed that the nation's overall welfare rested on a prosperous agriculture. No one had stated the principle more clearly than urbanite Theodore Roosevelt. In 1908 he wrote: "No nation has ever achieved permanent greatness unless this greatness was based on the well-being of the great farmer class, the men who live on the soil; for it is upon their welfare, material and moral, that the welfare of the nation ultimately rests." If farmers could capitalize on Americans' emotional attachment to the land and things rural, and on the belief that industrial greatness rested on agricultural prosperity, then this new minority might, with proper organization, achieve some kind of parity in the economy. In the early 1920s farmers began the most intense and best organized campaign ever undertaken to achieve equality for agriculture.

The Minority Fights Back

The Rise of Farmer Political Power

III

THE SHARP FARM DEPRESSION THAT BEGAN IN late 1920 had at least two positive effects for agriculture. It raised the political consciousness of many farmers, and it emphasized the need for effective organization. Historically, farmers were no strangers to organized effort. They had formed political parties, cooperatives, commodity groups, and a variety of farm organizations such as the Grange, but none of these had really succeeded in defending agriculture's interests. The post–World War I depression convinced more and more farmers and their spokesmen that they must be able to exert political pressure on Congress. In practical terms this meant organizing a strong farm lobby. Whatever farmers had done in the past to promote their interests in Washington seemed naive and unsophisticated compared to the activities of business and labor groups. This situation, however, was about to change.

For a variety of reasons farm influence in Washington reached a new high in the early 1920s. Indeed, modern agricultural lobbying dates from that time. Part of the increased power of farmers came from congressional sympathy for agriculture, which was suffering from a very severe depression. More important, however, was the fact that for the first time in American history farmers organized a truly national effort in the nation's capital.

The National Grange's history extended back more than a half century, but it was not until January 1919 that it established a full-time legislative representative in Washington. At that time the Grange had about a half million members. In addition, the National Farmers Union, first organized in the South in 1902, had expanded into the Middle West and the Great Plains states by the end of World War I and claimed some 140,000 members. This organization worked closely with the National Board of Farm Organizations, which was organized in 1917 and had a Washington office.

The most significant new element in the picture of developing agricultural power was the formation of the American Farm Bureau Federation. Growing out of the county farm bureau system, a group of state farm bureaus met in Chicago in November 1919 to create a national organization of farmers. Delegates representing more than 300,000 members from 28 states approved the new AFBF in March 1920. From the beginning, the Farm Bureau sought to improve the business aspects of farming, including better "marketing, transportation and distribution of farm products," and to develop a national farm policy. James R. Howard, an Iowa farmer and president of the Iowa Farm Bureau, became the organization's first president. One of the first acts of Howard and his executive committee was to appoint Gray Silver, a former state senator and fruit grower from West Virginia, as the AFBF's representative in Washington. While Silver said that he would not engage in "a lobbying campaign" and would confine his work to congressional education on agricultural problems, he soon became one of the best known and most effective lobbyists on Capitol Hill. The various farm groups that had established representation in Washington by 1920 did not work closely together—indeed, they were often at odds—but their total influence on legislative matters soon became evident.[1]

As the farm depression intensified and constituent complaints flooded lawmakers, another important development occurred. On May 9, 1921, Senator W. S. Kenyon of Iowa called together a group of his colleagues to discuss the agricultural situation. It is significant that this meeting was held in the AFBF office of Gray Silver. Out of that gathering, attended by six Republican and six Democratic senators, there emerged a loosely knit bipartisan group that became known as the farm bloc. Additional senators, mainly from the Midwest and the South, gradually identified with the bloc. In the House a group also began to devote special attention to farm problems.[2]

While farm bloc members did not agree on the means, there was general consensus that the purpose of the bloc's nonpartisan political activity was to seek special legislation to improve conditions among rural constituents. Many congressmen and senators believed this was essential, not only to help individual producers, but to guarantee national welfare. Senator Arthur Capper of Kansas, who soon became the recognized leader of the farm bloc, explained that "national prosperity is dependent primarily upon agricultural prosperity" and that without good times on the farm "the Nation cannot have a continued growth and development." Here was an expression of that deep strain of agricultural fundamentalism that resided in American thinking—the idea that agricultural welfare was synonymous with national well-being. Historically, Capper contended,

Americans had professed great faith "in the man on the land," but recently they had "developed an apathy toward the real needs of agriculture."[3] Something needed to be done to right this situation. However, Capper's justification for the farm bloc indicated a basic contradiction in American thinking. The rhetoric and sentiment of Americans was strongly on the side of farmers and farming, but peoples' interests and actions were aligning them more and more with the nation's industrial and commercial forces. Capper and other agrarians hoped to contain if not reverse this tide.

Another strong supporter for farmers in Washington after March 4, 1921, was Secretary of Agriculture Henry C. Wallace. Member of a prominent Iowa family and publisher of *Wallace's Farmer* in Des Moines, Wallace accepted the secretaryship because he believed that he could materially help farmers from that position. An outspoken champion of agriculture, Wallace saw himself as the special representative of farmers in Washington. He applauded the activities of the farm bloc and used the influence of his office to support specific farm-relief measures.

Leadership in the House and Senate also responded to farm demands in 1920 and 1921. In May 1919, Gilbert N. Haugen became chairman of the House Committee on Agriculture. A member of Congress from Iowa's rich northeastern district since 1899, Haugen was absolutely convinced that national welfare depended on farm prosperity. He drove his committee hard as it considered scores of legislative proposals after 1920. In 1921 George W. Norris assumed the chairmanship of the Senate Committee on Agriculture and Forestry, but Norris later shifted this job to Senator Charles L. McNary of Oregon. The House and Senate agriculture committees continued to be led by dedicated and effective legislators who had deep sympathy for farmers.

By 1921 agricultural interests had reacted vigorously to their minority position and to the hardships plaguing their industry. By establishing a strong political base in Washington, farmers indicated that they were determined to defend themselves. The sources of their strength were the AFBF and other farm organizations that had either national or regional constituencies, a strong Secretary of Agriculture, vigorous leadership of the agriculture committees in Congress, and the bipartisan farm bloc. Many who were unaccustomed to seeing farmers exert any effective political power decried the dangers posed by organized agriculture. The cry turned shrill in July 1921, when farm interests in Congress turned back a move to adjourn. Agricultural spokesmen insisted that the Senate stay in session and work on farm-relief legislation, despite the oppressive

Washington heat. Even though farmers may have had more power than previously, critics of farmer politics greatly exaggerated what was happening in Washington. It was certainly not true, as the *New York Times* reported on October 3, 1921, that the farm group had "obtained everything it demanded" in the current Congress.

Nevertheless, between 1921 and 1923 Congress passed what to many must have seemed like a flood of agricultural bills. Among the most important were the Packers and Stockyards Act, the Futures Trading Act, the Emergency Agricultural Credits Act, and two amendments to the Federal Farm Loan Act, all in 1921, the Capper-Volstead Cooperative Marketing Act in 1922, and the Intermediate Credits Act in 1923.[4] Farm interests lost an effort to place facilities at Muscle Shoals on the Tennessee River in northern Alabama under federal control in order to produce cheaper fertilizer, and to create a government corporation that would finance agricultural exports, but many observers recognized a new degree of agricultural political power. The *New York Times* bemoaned the fact that a Republican Congress could not achieve its major tax and tariff goals because of opposition by the bipartisan "farmer group." Congressman Martin C. Ansorge of New York became so agitated at the farm bloc that he introduced a bill making it illegal and punishable by a $5,000 fine for any congressman or senator to belong to any legislative bloc.[5]

Despite a spate of farm legislation in the early 1920s, divisions within agriculture were as deep as ever. The major farm organizations with strong support in the Midwest favored the protective tariff, but southern cotton producers, who sold a large portion of their crop abroad, opposed protection and discounted the home market argument. Some farmers believed that the formation of cooperative marketing and purchasing associations would solve farmers' price problems. Although there were examples of successful cooperatives, there were too many cooperative failures to produce majority support for this approach. Many farmers and their spokesmen urged an inflationary monetary policy, while still others demanded that the government guarantee prices that would equal the cost of production. If legislative activity on behalf of farmers between 1921 and 1923 seemed to reflect a high degree of unity and power among farmers and their representatives in Washington, this was more image than fact. The wide variety of interests among farmers and the many different ideas as to how those interests could best be served continued strong. Legislative proposals ranged all the way from opposing any kind of federal assistance for agriculture to insisting that farm prices should be established by legislative fiat in

Washington. Even the Farm Bureau, the only really national farm organization, experienced serious internal divisions over farm policy in the 1920s.

While the debate over how to alleviate farm ills continued, a small group of businessmen, farm organization leaders, and officials in the United States Department of Agriculture were working toward a new approach to farmers' problems. The leader of this group was George N. Peek, president of the Moline Plow Company from 1919 to 1924. By late 1921 Peek and his associate Hugh Johnson had developed a farm-relief plan that they called "equality for agriculture." Since it was surpluses that drove farm prices down, Peek suggested segregating the surplus of basic crops and letting the price of the domestically consumed portion rise behind a tariff wall. The price to be sought was one that would give the product the same purchasing power that it had in the prewar period from 1909 to 1914, a time when the exchange between farm and nonfarm prices was considered fair to farmers. This was the goal of price equality, or what eventually came to be known as parity.

Under Peek's plan, the surplus part of the crop would be sold abroad at world prices, thus eliminating the pressure on domestic markets. The program would be administered by a federal agricultural export corporation, which would sell the price-depressing surpluses overseas. Losses incurred on foreign sales would be recouped by placing a tax, called an equalization fee, on each bushel or unit of the commodity sold by producers. This, it was argued, would provide a fund to cover any export losses but would still leave the producer with an increased income because of higher prices received for that portion of the crop sold at home. The proposal sought to establish a two-price system and make the tariff effective on basic farm commodities. In January 1924, the principles advanced by Peek were introduced as the McNary-Haugen farm-relief bill.

The Peek plan appealed to those who did not believe that the traditional remedies—inflation, more credit, cooperatives, and regulation of monopolies—would significantly help agriculture. As farmers had become more highly commercialized and less self-sufficient, their main problem was what Gilbert Haugen called "this inequality in prices between agricultural commodities and other commodities." Low prices for farm products in relation to the cost of manufactured goods and services prevented farmers from getting fair returns on their capital and labor. Thus the idea of being treated equally or achieving parity in the economy was highly attractive. Who could oppose the concept of fairness to a major group such as farmers?

One of the main reasons individualistic farmers had not fared better was their poor bargaining position with organized groups in the economy. Supporters of the Peek proposal said that surplus-control legislation would give farmers an opportunity to unite through a government corporation and increase their bargaining power with nonfarm elements in the economy. It was not a matter of farmers seeking special privilege, they said, but of achieving a position that would assure them equal treatment.

Another argument for the McNary-Haugen legislation went to the very heart of agrarian thought. The plan's backers insisted that without good times on the farm the country as a whole could not long enjoy prosperity. Agriculture always had been, and still was, the nation's basic industry. If that fundamental industry suffered, so would all the people. Agricultural spokesmen believed this with religious fervor.

The idea that a government corporation should be employed to raise the prices of certain basic farm commodities found strong opposition in many quarters. President Calvin Coolidge said in December 1923 that "no plan for government fixing of prices, no resort to the public treasury will be of any permanent value in establishing agriculture." Coolidge believed that farmers must work out their own salvation, or, at best, join together in voluntary cooperative associations. Most of Coolidge's cabinet, notably Secretary of Commerce Herbert Hoover and Secretary of the Treasury Andrew Mellon, opposed any legislation that would lift farm prices. After the death of Secretary of Agriculture Wallace in October 1924, the new Agriculture Secretary, William Jardine, also joined the critics. Likewise, business interests, the metropolitan press, economists, consumer groups, and even many farmers opposed the McNary-Haugen legislation. The major farm organizations were divided on the question. As a result of this formidable opposition and the lack of sufficient organization by supporters, the House defeated the McNary-Haugen bill in June 1924.

It was clear to Peek and others after this first battle for surplus-control legislation that if farmers were ever to obtain legislation that would assure them equality of treatment, they must get better organized. Farmers simply could not count on the general goodwill of national lawmakers to meet their needs. Consequently, on July 11, 1924, about 150 farmers and agriculture leaders met at St. Paul, Minnesota, to consider forming an organization that could achieve enough political influence to force helpful farm legislation through Congress. Besides Peek, Gray Silver, and other experienced lobbyists, representatives of the three largest farm organizations, the Grange, the AFBF, and the Farmers Union, were present. The group quickly decided to form the American Council of Agriculture, whose chief object was "to make it possible for the existing agricultural organizations of whatever character to speak with one voice through a united leadership wherever and whenever the general well being of agriculture is concerned." In practical terms this meant gaining equality for agriculture with industry and labor through passage of the McNary-Haugen bill.

To head the new American Council of Agriculture, the delegates chose George Peek. He was an ideal choice. Peek had the time and talent, the financial independence, and a deep commitment to the cause of agriculture. He had no association with any of the major farm organizations and thus avoided the suspicion and jealousy that might have developed if the leadership had come from any one of those groups. Peek had many contacts in the business community, including financier Bernard Baruch, which, if mobilized on behalf of farmers, could greatly strengthen any farm lobby. Moreover, despite his long business career he was a strong agricultural fundamentalist. Peek believed that the welfare of farmers and that of the nation as a whole was absolutely interdependent.

Peek soon established an office in Washington and set out to mobilize an effective farm lobby. This would not be an easy task. By 1924 the bipartisan farm bloc had disintegrated under the pressures of party politics and the improvement of some farm prices. Even more important was the fact that southern cotton producers had never been enamored by the McNary-Haugen bill. They had not been convinced that it would help cotton, a large portion of which was exported. The main problem facing the new farm lobby was to convince southern congressmen and senators that surplus-control legislation would be beneficial to their section. The need was for political unity between the agricultural Midwest and the South, a marriage of the interests of corn, wheat, and cotton.

Beginning in late 1924, Peek and his associates lobbied hard for farm-relief legislation that would give farmers what they called ratio prices for basic products. They worked with the regular farm organizations, organized additional groups, such as the Corn Belt Committee of Twenty-Two, testified before congressional committees, distributed information, stirred up grass-roots opinion, buttonholed congressmen and senators, and raised money. The Coolidge administration, however, was a formidable foe. By frontal attacks, as well as through diversionary tactics, the Coolidge forces thwarted passage of any surplus-control legislation in 1925 and 1926.

Farm lobbyists also sought strength by continuing their appeal to the basic philosophical belief that agriculture and farming were at the foundation of American greatness. For those accepting that idea, it followed that any decline in agriculture would be fatal to both national economic welfare and political stability. Frank W. Murphy, a Minnesota farm leader, raised his listeners to a high emotional pitch when he told delegates at the St. Paul meeting: "If you would understand the soul of America, if you would catch the fire of that great spiritual force that keeps the flame of patriotism burning in this land of ours—if you would know the source from which flows that great current of fine purposes, high resolves, courageous, clean, Christian citizenship, visit the shrines of American farm homes, and there you will find it all." Congressman Haugen declared that without prosperity on the farms, "just as sure as the sun rises in the East and sets in the West . . . our factories and mills and our banks would crumble to pieces" and the nation's "magnificent institutions would materially suffer." Such statements and declarations on the economic and political importance of agriculture became a regular litany at farm meetings and on the floor of the House and the Senate.

The thing that so infuriated farmers was the belief that the federal government was openly aligning with business and industry, and even labor, at the expense of agriculture. Treasury Secretary Andrew Mellon and other opponents of the McNary-Haugen bill gave credence to this belief. Mellon opposed surplus-control legislation because, he wrote, it would "increase the cost of living to every consumer of five basic agricultural commodities." As Mellon saw it, "we shall have the unusual spectacle of the American consuming public paying a bonus to the producers of five major agricultural commodities, with a resulting decrease in the purchasing power of wages." At the same time, he continued, cheap exports of American foodstuffs would subsidize foreign consumers, lower industrial costs overseas, and permit foreign manufacturers to undersell American producers in world markets.[6]

The Mellon position seemed to confirm the belief of farm leaders that the Coolidge administration was bent on enhancing industry while agriculture languished. Peek declared that "no one can hear or read the utterances of the leading administration's spokesmen without coming to the conclusion that they are determined to have cheap foodstuffs and raw materials for industry regardless of what happens to the farmers." In November 1927, a writer for the *National Farm News* (Washington) declared that the East wanted cheap food "regardless of what happens to the producers." But more than that, poverty on the farm would hasten the cityward movement, add to the labor supply, and permit industrialists "to name their own wage scales and working rules." The basic question of the time, Peek said, was "shall we industrialize America at the expense of agriculture or shall we retain the American tradition of the independent, landowning farmer? It is the most profound question the nation has faced since the Civil War." Peek had raised a basic question, but far too late for any action to be taken to reverse the trend.

Farmers viewed the McNary-Haugen bill as both symbol and substance. One popular writer explained the situation by saying: "it [McNary-Haugen] is the test that determines whether special privilege is to be for urban man and special discrimination for the farmer." The question was not so much whether the bill would help agriculture, but whether Congress and the President were for or against the demand for equality when largess was dispensed from Washington. Some opponents of the McNary-Haugen bill urged farmers to reduce surpluses and solve their price problems by producing only enough for domestic demand. Most farm leaders responded angrily to that suggestion. To restrict farm exports while expanding the sale of manufactured goods overseas, they charged, would severely hurt farmers. If such policies prevailed, columnist Mark Sullivan observed, "the farmer's economic status, and his social status, will be that of gardener to an immense manufacturing and business community." Sullivan agreed with the farmers who believed that such a policy would lead to a "definite subordination of farming to other industries."[7]

By early 1927 the farm lobby led by Peek had organized enough political power to force a revised McNary-Haugen bill through Congress. Southerners were brought into the campaign as a result of a huge cotton crop in 1926, which caused a severe drop in prices that fall. Leaders of the American Farm Bureau Federation had successfully developed a unity of interest between the organization's southern and midwest groups. Senators and congressmen in the South who had previously been cool to McNary-Haugenism now lined up behind the legislation that provided special help for marketing cotton. Both houses passed the measure in February 1927, under what the editor of the Wichita (Kansas) *Beacon* called "the bludgeoning of one of the most persistent and skillful lobbies ever seen in Washington." There is no doubt but that farm groups had exerted intensive pressure on wavering legislators. One House member said he had to vote for the bill "because the crowd at home are on my trail." Senator James Watson of Indiana, who had opposed the bill earlier, voted for it in 1927. When a friend asked, "Jim, how come," the Senator replied: "Well, you know there comes a time in the life of every politician when he must rise above principle."[8]

There was little surprise when President Coolidge vetoed the farm bill in a stinging message. He opposed the legislation because, he said, it "involved government price fixing," ran counter to "the spirit of our institutions," would stimulate production, breed a "cancerous bureaucracy," and was unconstitutional. Unable to override the President's veto, the farm lobby realized that to be successful it must somehow revise the legislation to meet Coolidge's objections. In the end this proved to be impossible and still retain the equalization fee which farm spokesmen considered the "heart and soul" of the bill. Both houses of Congress passed the McNary-Haugen bill again in the spring of 1928, but Coolidge killed the measure with a second veto. Writing in the Baltimore *Sun*, Frank R. Kent said that the President's message "first hit the McNary-Haugen bill squarely in the nose. Then it kicked it full of holes. Finally it swept it up in the corner and struck a match."[9] By that time farm conditions had improved considerably, but the farm pressure groups wanted their program as a matter of principle.

President Coolidge's strong opposition to the McNary-Haugen bills indicated that no surplus-control legislation could be passed until a change occurred in the White House. The prospect of Hoover's nomination and election was not encouraging to the farm groups because he had been a leading opponent of McNary-Haugenism. George Peek and other Republican farm leaders favored the nomination of Frank O. Lowden, former governor of Illinois, who had supported their farm fight. Julien N. Friant, a farm leader in Missouri, agreed with Peek's position and wrote that for farmers to turn their welfare over to Hoover, who had opposed them for eight years, was "as foolish as for a girl to marry a drunkard with the hopes of reforming him."[10] Hoover, however, easily won the Republican nomination. Peek then urged the friends of agriculture to support the Democratic candidate, but wet, Catholic Alfred E. Smith from the sidewalks of New York had little appeal to farmers. While Peek hoped to make the presidential election of 1928 a referendum on the farm-relief issue, higher agricultural prices and Smith's unattractiveness to farm voters made this impossible. Hoover's overwhelming victory all but killed McNary-Haugenism and sent farm leaders home to wait for some new turn of national events.

Despite the defeat of effective farm-relief legislation in the 1920s, much more had been gained than seemed obvious at the time. Through lobbying activities, farmers and their spokesmen had gained valuable political experience. Farmers had increased their knowledge and sophistication of the political process and of what it took for them to make the system respond to their needs. They discovered that with proper effort and organization they could compete in Congress with other special interest groups. Also the McNary-Haugen campaign established the concept of economic equality or parity for farmers. This principle became the basis for most subsequent agricultural legislation dealing with prices. Moreover, people accepted the idea for the first time that the federal government must play a major part in achieving the goal of farm parity. Producers understood clearly that the free market would not assure them profitable prices and a decent income so long as they had to deal with elements in the nonfarm economy that set prices by monopoly power or by administrative edict.

President Hoover supported the Agricultural Marketing Act enacted by Congress in 1929. This law provided federal support for agricultural cooperatives, which were supposed to keep price-depressing surpluses off the market through orderly marketing. The measure also permitted the organization of federally financed stabilization corporations to maintain the prices of selected farm commodities in case the cooperatives failed. Most McNary-Haugenites predicted that the law would fail, but they did not actively oppose it. In the political climate of early 1929, it was clear that no other type of farm relief could be enacted. By 1931 the Hoover plan had failed.

The stock market crash of October 1929 dramatized serious weaknesses in the American economy and set in motion events that led to one of the nation's longest and most severe depressions. All aspects of the economy were suffering by 1931, as plants closed, banks went broke, business slowed, and unemployment lines lengthened. By 1932 nearly one-fourth of the labor force were without jobs. Some people considered living on a farm preferable to urban unemployment, and thousands of individuals who had earlier sought their fortunes in the city drifted back to rural communities and the farm of their youth. But anyone who returned to the farm in 1931 or 1932 found depressing conditions. Wheat sold for as little as 25 cents a bushel in 1932, cotton brought less than a nickel a pound, and livestock hardly paid the cost of marketing. The situation was aptly illustrated by the colorful but apocryphal story about the farmer who sent several head of sheep to market and received a bill from the commission firm for $5 with the explanation that the sale price had not paid the selling costs. The farmer responded by shipping more sheep! From 1929 to 1932 total net income to farm operators slid from about $6.1 billion to approximately $2 billion.

Farmers were no strangers to hard times. Even during the 1920s, when other elements of the economy were fairly prosperous, many farmers experienced economic difficulties. That is what the farm fight had been all about. Long before the Great Depression began, many midwesterners and farmers on the Great Plains had lost their farms through foreclosure, while tens of thousands of landless sharecroppers and tenants in the South wallowed in extreme poverty. But now conditions created new and shocking levels of want and destitution beyond anything previously known. By 1932 the average net income of people living on farms had dropped to $74 a year.[11]

Farmers described their pitiful condition in letters to state and federal officials. An Indiana farm wife wrote the Secretary of Agriculture in 1930 explaining her family's situation. Even though her husband was a graduate of Purdue University and they had managed to acquire a 250-acre farm, they were about to lose everything. They were falling behind, she said, even though they raised what they ate and were "worked to death with no income, no leisure, no pleasure [and] no hope of anything better." She had talked to some farm women who were so depressed by hardship that they wished for the end of the world. "We are a sick and sorry people," she wrote, "the very country is dejected and forlorn. My nearest neighbor has turned bootlegger, I can smell the mash brewing in his still."

A Missouri farmer said that in his thirty-nine years of farming, conditions had never been so bad, while a North Dakota farm woman wrote in July 1931 that her family had no money and was losing its farm. "My husband and I are losing everything," she wrote, despite the fact that they had worked from 4:30 in the morning until 9:30 at night. Another farm wife wrote: "it is just terrible. Eggs and cream are not worth anything. Hogs have gone down until they are not bringing anything either. A lot of our prices are fixed. Interest, taxes, insurance. . . ."[12]

The Great Depression seemed to confirm the argument advanced by farmers and their supporters that the nation could not prosper unless farm income and purchasing power were strong. Although business and industry had boomed for a time in the 1920s, despite low returns to agriculture, farm spokesmen argued that permanent prosperity could not be achieved if farmers were excluded. By 1931 and 1932, as the grip of depression tightened, an increasing number of Americans were coming to this view, and insisted that strong and positive action be taken to restore the agricultural sector.

While farm groups had been preaching this doctrine for years, business and labor now joined in paying tribute to agriculture's economic importance. In August 1932, an article entitled "Agriculture is the Foundation of Manufacture and Commerce" appeared in the *Manufacturers Record.* Author Charles D. Bohannan said that he took his title from the USDA seal which "expresses a truth that has peculiar significance in our present economic situation." The Depression could have been averted or at least minimized if people had not forgotten that whatever affects the buying power of farmers affects the entire economy. "This fundamental fact," he continued, "needs to be driven home to all." Bohannan concluded: "The evidence unmistakenly points to the conclusion that economic soundness for agriculture is a *sine qua non* for the welfare and prosperity of all the other elements in our national economic and social structure." Other business and labor magazines advanced this same argument. Writers for farm publications expressed the same opinion and often added: "We told you so."[13]

As the Depression worsened, farm groups concentrated on developing legislative proposals that they hoped would lift farmers out of the economic mire. Equally or more important was their desire to nominate and elect a president who would be sympathetic to farm needs. As it turned out, farmers were successful on both counts.

While some farm leaders continued to advocate McNary-Haugenism, there was no practical chance to enact such legislation in 1931 or 1932. Moreover, there was no likelihood that the McNary-Haugen approach to farm relief could have succeeded under world conditions at that time. The lack of domestic demand and the drop in exports had left American farmers with huge price-depressing surpluses that no export corporation could have handled without exorbitant sums of money to cover losses. In light of the burgeoning surpluses and the low prices, a small group of economists, businessmen, and farm leaders began approaching the agricultural dilemma from a different angle. The heart of their thinking was to raise prices by restricting output. Rather than trying to dispose of surpluses after they were produced, these people reasoned that it would make more sense not to produce excess commodities in the first place.

Although the idea of production control went back to the 1630s when Virginia and Maryland tobacco growers advocated restricting output to raise prices, the great majority of American farmers opposed planned restrictions on production. Most farmers viewed their main purpose to be that of producing as much and as efficiently as they could with the land, labor, and capital at their disposal. The main programs of the agricultural colleges, the extension service, and the USDA were designed to encourage greater output. At the same time, however, farmers saw that manufacturers had much greater control over their prices because they could regulate production closely to the prospective market. If supply control were good for business, why would it not be equally beneficial for farmers? An example of the differences between the market for agricultural and that for other commodities can be shown by pointing out that between 1929 and 1933 farm prices dropped 63 percent while the prices of agricultural implements and motor vehicles declined only 6 and 16 percent respectively. These industries reduced their production 80 percent and maintained relatively high price levels. Considering prices in the years 1909 to 1913 as equaling an index of 100, by 1932 the purchasing power of farm products was only 58.

The first stage in the move toward a program of restricted production came with the development of the domestic allotment plan. While the principles had been suggested earlier, it was M. L. Wilson, a professor of agricultural economics at Montana State College, who in 1930 began refining and promoting the idea. He won the support of such prominent individuals as Henry A. Wallace, Henry I. Harriman (who became president of the United States Chamber of Commerce in 1932, Senator Peter Norbeck of South Dakota, and many others. The domestic allotment plan went through several stages, but the emphasis was on giving farmers direct price incentives to reduce production of basic crops. In its final form the plan called for distributing cash benefit payments to farmers who signed a contract limiting their cultivated acreage. The payment was to approximate the amount of the tariff on that portion of the crop used in domestic consumption—the volume grown under a producer's domestic allotment. Money for the benefit payments to farmers was to come from taxes on processors of agricultural commodities.

Wilson and his backers had not shifted the goal of farm-relief legislation, but they had greatly changed the means to achieve it. Equality for agriculture was to be obtained by balancing production and consumption. The emphasis was upon production, rather than marketing as had been the case with the McNary-Haugen and other farm bills of the 1920s. In taking this approach, Wilson said they were "making history right now," an assessment with which many observers later agreed.[14]

The idea of acreage restriction did not set well with most farm leaders or their constituents. In 1932 officials of the American Farm Bureau Federation damned the concept. In September 1932, when W. R. Ronald, editor of the Mitchell (South Dakota) *Daily Republic* and a key propagandist for the domestic allotment plan, was trying to sell the idea to William H. Settle, Indiana's Farm Bureau leader,

Settle disgustedly exclaimed: "What is the idea of wasting our time talking about a blankety blank idiotic farm plan when everybody knows we are for the McNary-Haugen bill." John A. Simpson of Oklahoma, militant president of the National Farmers Union, also expressed strong opposition to the domestic allotment scheme. He called it a "United States Chamber of Commerce scheme to muddy the waters." George Peek was especially critical of reducing production "at the expense of normal exports." He saw this as another move by industry to restrict agriculture while maintaining its own advantage in foreign markets.

It made little difference, however, what new ideas or approaches might be proposed to solve agricultural problems unless there was a change in the White House. Hoover would brook no nontraditional ideas on how to deal with agriculture. One of the most important developments in the fight for economic equality for agriculture was the nomination and election of Franklin D. Roosevelt.

One thing was clear enough about the Democratic nominee. He was deeply sympathetic with the plight of farmers and held to the main tenets of agrarianism. Roosevelt boasted of his connection with the land. When someone once referred to his Hudson River estate, he responded, "call it by its right name, a farm. I don't like estates and I do like farms." Roosevelt had supported a back-to-the-land movement in 1930 and 1931, because, he said, in small communities dotted with industries workers would find "a more wholesome life, away from the grime and misery of city slums." Where people could have contact with the earth and with nature, Roosevelt continued, they would find "an opportunity for permanency of abode, a chance to establish a real home in the traditional American sense." As Frank Freidel, Roosevelt's biographer, has written, Roosevelt had a "romantic faith in the Jeffersonian ideal of the independent yeoman living in bucolic abundance." While Roosevelt had no firm convictions on how best to solve the farm problem, he had an open mind and was willing to try different and unconventional methods.

The Democratic platform on agriculture called for "enactment of every constitutional measure that will aid the farmers to receive for their basic commodities prices in excess of cost." Roosevelt, however, had not made up his mind at the time of the convention on how best to attack the agricultural problem. During the late summer and fall of 1932 there was a struggle among farm spokesmen to win Roosevelt's support for their particular position. John Simpson believed that the Democratic platform committed Roosevelt to the Farmers Union program, which called for a guarantee of cost of production for farm products. Those favoring McNary-Haugenism urged Roosevelt to back that program, one which they said farmers supported and understood. Supporters of the domestic allotment plan, however, gradually won Roosevelt's endorsement.

During the fall of 1932, leaders of the major farm organizations held several meetings in an effort to agree on a national agricultural policy. Edward A. O'Neal, who had become president of the American Farm Bureau Federation in 1931, was the moving force behind several of these meetings. O'Neal, a northern Alabama planter, had been influential in bringing cotton support behind McNary-Haugenism and in forging a southern-midwestern coalition in the mid-twenties. His move to the national organization's presidency was a timely and even lucky event for the political power of farmers. As a Democrat, he headed the nation's largest farm organization just as the southern Democratic leaders were assuming control over legislative matters. Moreover, he had proven talents as a mediator of conflicting agricultural interests.[15] It was not known just what plan the American Farm Bureau Federation would support, until December when O'Neal announced his support of the domestic allotment proposal. What had changed O'Neal's mind? Roosevelt told O'Neal that as President he would not endorse any program that did not include acreage control. According to M. L. Wilson, O'Neal jumped on the bandwagon when he learned which way the new administration intended to go. "Nothing so delights the head of an organization in any field as to find out in advance," Wilson related, "what is going to be done by the administration in Washington and, if at all acceptable, to get in line for it, because it puts him in a strong position." By the end of 1932 and the early weeks of 1933 most farm groups had come to support production control and benefit payments, principles that were later incorporated in the Agricultural Adjustment Act. Among the leaders of the three major farm organizations, only the Farmers Union held out for another approach. John Simpson sharply attacked the idea of acreage restriction and continued to insist that the federal government guarantee farmers prices equal to the cost of production.

Meanwhile, because of the strain of the Depression, the patience of some farmers had become completely exhausted. Under the leadership of the fiery Milo Reno, leader in the Iowa Farmers Union, producers in Iowa and Minnesota were turning to direct action to get improved prices and to protect their lands and chattels from foreclosure. A group of some two thousand farmers met in Des Moines on May 3, 1932, and formed the Farmers' Holiday Association. Reno called for a strike to begin on August 8. Farmers set up picket lines outside a number of towns in Iowa and Nebraska and declared that no produce would be permitted to reach the markets. In some places pickets dumped milk in roadside ditches, battered vehicles with clubs, and spread spikes on highways to deter trucks. While the roads were cleared within a few days, ugly incidents continued to occur. Reno and other leaders of the Holiday movement urged farmers to withhold produce from market, but advised no further picketing. During the fall of 1932 the situation was relatively quiet as farmers waited to see what President-elect Roosevelt might do.

Early in 1933, however, more lawlessness occurred. This time farmers used force to stop foreclosures on land and personal property. Unable to pay debts with 35-cent wheat, 20-cent corn and 3-cent hogs, angry farmers showed up at auction sales and bid in their neighbor's property for a few cents and returned it to the original owner. Any potential outside bidders were threatened with physical violence if they entered a bid. Late in April, 1933, a group of irate Iowa farmers dragged a judge from his bench, threatened to hang him, and mistreated him shamefully before leaving him bruised and battered. Congressman Guy Gillette of Iowa remarked: "I sympathize fully with the farmers and realize they are driven to desperation by conditions over which they have no control." It took a declaration of martial law to restore order in some Iowa counties. Despite lawlessness and radical rhetoric, these farmers did not want to endanger the democratic-capitalistic system. They were unwilling, however, to stand by and see outside forces take away property and homes for which the farmers had worked a lifetime. They wanted what they considered their share of the fruits of the system.[16]

While the activities of the Farmers' Holiday Association brought widespread attention to farm problems, it did not take any grassroots protest to convince either President Roosevelt or Congress early in 1933 that legislation to raise farm prices must have high priority. Not only did something need to be done for farmers as a group, but many people believed that farm incomes must be raised before general prosperity could be restored. Thus, in passing agricultural legislation Congress would be helping farmers and the entire economy at the same time.

On March 10, 1933, the day after Congress met in special session, Secretary of Agriculture Henry A. Wallace called farm leaders to a meeting in Washington to consider the principles of new farm legislation. Within two days they had agreed on the main outlines of a farm bill which was introduced a few days later. By that time the main purpose and principles of new farm legislation had been generally accepted by farm spokesmen. The goal of New Deal farm policy was to give agricultural commodities the same "equality of purchasing power" as they had enjoyed in the base period 1909–14. The means of achieving this equality, or parity price, was to balance "production and consumption" by enticing farmers voluntarily to produce less. The Secretary of Agriculture was given broad powers to enter into contracts with individual producers of seven basic commodities and pay them for reducing their output. Money for these payments was to come from taxes on processors of farm commodities. The amount of the payment was to be the difference between the farm price of the commodity and the "fair exchange value." However, the Secretary of Agriculture had considerable latitude in administering the law and was not required to offer benefit payments that equaled full, or 100 percent, parity prices.

Other important sections of the Agricultural Adjustment Act, which was approved on May 12, 1933, were the provisions for marketing agreements, the Thomas inflation amendment, mortgage relief, and protection for consumers. George Peek, who really never accepted the principle of acreage restriction, believed that the United States could sell more farm commodities abroad if it would enter into special arrangements with foreign countries, including subsidizing exports. Senator Elmer Thomas of Oklahoma, responding to the growing demand for inflation, added an amendment that permitted the President to inflate the currency. Moreover, recognizing that processors would pass the tax along to consumers, and being sensitive to any rise in food prices in a period of high unemployment, the lawmakers included a consumers' protection clause. This section restricted adjusting farm production beyond the point where "the percentage of the consumers' retail expenditures for agricultural commodities . . . which is returned to the farmers" would go above the level of the prewar period. The purpose was to keep farm prices from rising above the parity figure and to keep food prices low.[17]

There was sharp argument over both the philosophy behind the new law and the practicality of its operation. While it passed by large margins in both houses of Congress, some critics attacked the plan of restricting production when people were hungry. They reversed the common theme that economic restoration depended on increasing farm purchasing power, and argued that farmers could best be helped by creating jobs for the unemployed. A New Jersey truck gardener wrote that "the only way possible to get money in the farmers' empty pocket will be to place twenty million jobless men back to work." Others scoffed at the idea that farm production could be regulated at all. As John Simpson, one of the bill's severest critics, said: "It is impossible. You would have to have God on your side to be sure that it would work." Besides opposing production control, Simpson continued to push his amendment, which would establish farm prices at the cost of production. There was a great deal of grass-roots support for this approach, which seemed like a simple and direct way to raise farm prices. One Texas farmer wrote Simpson: "We are with you even if we are living on beans and sow-bosom." However, strong administration opposition defeated Simpson's attempt to add a cost-of-production amendment to the bill.

Other opponents fought the Agricultural Adjustment Act because they considered it radical and un-American. Secretary Wallace was accused of holding Czarist powers over farmers, while one congressman said the farm program would "sovietize" American agriculture. The extension of government powers and responsibility envisioned in the legislation was totally unacceptable to many people. As Congressman U.S. Guyer of Kansas expressed it: "This measure puts a policeman on every farm, an inspector at every crossroad, and a Government agent in every back yard."[18]

Once the law was passed, the Agricultural Adjustment Administration, headed by George Peek, set out to bring benefits to farmers as quickly as possible. Since most crops had already been planted, it posed a problem of how to pay farmers for reducing acreage. But the AAA actually did what only a short time earlier would have been unthinkable. It contracted with about 500,000 cotton farmers to plow under more than 10 million acres of the crop in return for benefit payments. While officials were agonizing over the possibility of also having to plow up wheat, unfavorable weather conditions in much of the wheat country made that unnecessary. Wheat farmers got payments in return for promising to reduce their acreage in 1934 and 1935. A little later, a program for corn-hog farmers called for cutting hog numbers by slaughtering young pigs and pregnant sows. The carcasses were used for food and fertilizer. Cries of anguish arose over the destruction of these little pigs, although critics never made it clear why it was worse to kill a small pig than a grown hog. By September 1933, the postal service was distributing hundreds of thousands of government checks to farm mailboxes, providing the first substantial amount of money many farmers had enjoyed for nearly two years. After observing the check-writing machine and having the purpose explained to him by USDA officials, a Russian visitor exclaimed: "Good Lord! This is a Revolution."[19] Many Americans agreed.

Passage and implementation of the Agricultural Adjustment Act expressed some basic assumptions in American political thought and action. First and most important was the nation's commitment to the parity principle for farmers. Congress had written into law the concept, if not the reality, of fairness to agriculture in the matter of price relationships. The legislation assumed that farmers as individuals were at a disadvantage in dealing with other economic groups and needed the centralizing power of government to place them in a stronger bargaining position. Furthermore, the law reflected a consensus that it was essential to restore the purchasing power of farmers before the country could emerge from the Depression. There was also a widespread belief that the AAA was needed to make the family farm attractive and profitable in order to maintain a stable population distribution between the cities and the countryside. Certainly there was abundant evidence that the majority of Americans still believed strongly in the agrarian tradition.

It was not clear at the outset how long the AAA would be needed. The law permitted the President to terminate the act whenever he considered that the agricultural emergency had ended. Once the law was on the books, however, it became what one authority has called "the nucleus of a long-run system of agricultural planning." The original act applied only to seven basic commodities—cotton, wheat, corn, hogs, tobacco, rice, and dairy products. Since the 1920s, farm relief advocates had believed that if the prices of these major commodities could be raised, most other crops would follow along. Gradually, however, under political pressure from farmers, other crops were added to the "basic" list. These included barley, rye, flax, peanuts, sugar beets, sugarcane, potatoes, grain, sorghum, and for a time, cattle. The addition of these crops to the list eligible for government benefits resulted from effective political trade-offs among congressmen and senators who represented different aspects of agriculture. Spokesmen for corn and wheat farmers discovered that if they expected continued support for their programs, part of the price was to vote to include such crops as peanuts and sugarcane.

By the middle 1930s, a powerful coalition of special farm interests had developed in Washington. The main coalition was between the Middle West and the South, but when other commodities such as sugar beets and potatoes were added to the favored list, agricultural political power extended much further. In a sense what developed was farm commodity power translated into political influence which guaranteed government benefit payments and price supports for the crops of the nation's most productive and influential farmers.

However important the adjustment of agricultural production and payment of cash benefits to cooperating producers may have been, it was only one aspect of a broad program of federal assistance extended to farmers in the 1930s. Title II of the Agricultural Adjustment Act provided for emergency farm mortgage relief. Tens of thousands of farmers avoided foreclosure through the liberal credit and refinancing arrangements that this law made possible. In June 1933, Congress passed the Farm Credit Act, which provided for a comprehensive federal agricultural credit system. By 1940 at least a million farmers had been helped by the various federal credit programs. Under strong pressure from rural congressmen and senators, some federal agencies provided credit to farmers in the form of subsidized interest rates. Furthermore, in 1934 Congress passed the Frazier-Lemke Farm Bankruptcy Act, which under certain circumstances permitted debt-ridden farmers a five-year moratorium against foreclosure. Although this law was declared unconstitutional, a revised measure was enacted in 1935 which stood the constitutional test.[20]

In October 1933, President Roosevelt established the Commodity Credit Corporation by executive order. The purpose of this agency was to make federal loans on properly stored wheat and cotton in order to keep price-depressing surpluses off the market. Only farmers who had signed acreage restriction contracts were eligible for these so-called nonrecourse loans. The farmer was fully protected. If prices went above the loan level, the borrower could sell the commodity, pay off the debt, and take his profit; if the price fell below the loan figure, the CCC took the commodity and the loss. The effect of these loans was to place a floor under the price of commodities it supported. While loans were made initially only on wheat and cotton, the program was soon extended to other crops.

The federal government also assisted farmers by helping them to conserve their land. Wind and water erosion, combined with poor farming and ranching practices, had destroyed or greatly reduced the fertility of millions of acres, especially throughout the South and the Great Plains. It was often found that poor land and poor farmers went hand in hand. To deal with problems of conservation, the Soil Erosion Service was set up in the Department of Interior in 1933. This agency surveyed land resources, encouraged planting of trees and grass, urged farmers to terrace their fields, and gave technical assistance. In 1935 the Soil Conservation Service in the Department of Agriculture took over the functions of the Soil Erosion Service. Under both programs farmers received government payments for implementing a wide variety of soil conservation practices.

Another federal program that was helpful to farmers, as well as to consumers, was the distribution of surplus agricultural commodities to needy citizens. The Federal Surplus Relief Corporation, which was replaced by the Federal Surplus Commodities Corporation in 1935, distributed millions of pounds of pork, butter, flour, and other items. While much of the food came from commodities under acreage or production control, the FSCC also entered the market to remove temporary surpluses of apples, citrus fruits, eggs, dried beans, and other products, and made them available to people on relief. For instance, in 1936 the FSCC purchased several hundred carloads of surplus grapefruit for distribution through relief channels. This program did much to improve the diets of many citizens and strengthened agriculture's image among poor consumers who found these surplus commodities little short of a godsend.[21]

Aspects of the Tennessee Valley Authority, established in 1933, also aided many distressed farmers in the Upper South. Ever since World War I farmers had been interested in using water power developed from the Tennessee River to produce cheap fertilizer. Senator George W. Norris of Nebraska, sometimes called the father of the TVA, as well as President Roosevelt, had much more ambitious plans for the region. They wanted to rebuild and improve the economy of the entire area. While production of electricity and flood control were important objectives, the TVA also worked with farmers to improve their agricultural practices, especially through better fertilization and soil conservation, and "to promote and encourage the fullest possible use of electric light and power on farms."

No government program of the 1930s, however, did so much to change rural life-styles as did the Rural Electrification Administration. Set up by executive order in 1935 and given congressional sanction the next year, the REA provided the needed support to electrify American farms. While private power companies had extended service to a few farm customers, rural business was not considered profitable. It was clear that if farmers were to wait on the expansion of private lines they would die in the dark. The REA granted loans at very low interest rates to rural electric cooperatives which built and managed electric service in rural communities. In 1934 about 744,000 farms had electricity, but the number reached 2,351,603 by 1941. This was nearly 40 percent of the farmers in the United States.

Despite a multipronged attack on agricultural problems by the federal government, the heart of the New Deal's program seemed to have suffered a crushing blow in January 1936. The Supreme Court declared the Agricultural Adjustment Act unconstitutional on the grounds that the Constitution did not authorize Congress to regulate farm production or to levy a processing tax to make benefit payments. Congress, however, had no intention of leaving farmers unprotected. Within a few weeks the lawmakers passed the Soil Conservation and Domestic Allotment Act. This law was designed to do the same thing that the AAA had done—namely, restrict acreage and make benefit payments to cooperating producers. But farmers ostensibly were not to be paid for cutting production, but for conserving their soil. Thus payments would be made to farmers who shifted from soil-depleting crops to soil-conserving crops. It just happened that the soil-depleting crops were those in surplus, such as wheat and cotton. The law also contained some provisions for additional conservation payments.

The most important change in this legislation was the goal to be achieved. The original AAA had called for giving specified farm commodities the same purchasing power as they had in the years 1909 to 1914. The new legislation forsook the principle of parity prices and substituted the objective of parity income. The law stated that the objective was to reestablish "the ratio between the purchasing power of the net income per person on farms and that of the income per person not on farms" that prevailed during the years 1909 to 1914. From the farmers' point of view, this was a superior objective, but, as will be seen, one seldom achieved.

Two years later Congress passed the Agricultural Adjustment Act of 1938, which became the basic law for future years. It continued the soil conservation features of the 1936 statute, and spelled out the policies for production control and benefit payments. The method for determining parity prices was modified and the goal of parity income for farmers was defined more clearly. A new feature added was the provision for marketing quotas. If the Secretary of Agriculture saw the prospect of excessive and price-depressing surpluses, he could limit commercial sales of a particular crop to each producer on a pro rata basis. If the farmer sold more than his quota, he was subject to a stiff tax on the transaction. Before quotas could be imposed, however, two-thirds of the producers of a particular crop had to vote in favor of the policy. Price supports were continued through the Commodity Credit Corporation. The level of support varied from 52 to 72 percent of parity, depending on the amount of production of a particular crop. The higher the output, the lower the loan rate. The idea behind this provision was to use a so-called flexible price-support system to discourage excess production. The 1938 law also authorized the Secretary of Agriculture to make direct parity payments to farmers under some conditions. The amount equaled the difference between the market price and parity price. Marketing agreements were continued. The 1938 AAA was a comprehensive piece of legislation which aimed at dealing with farm problems on several fronts.

Commercial farmers were the main beneficiaries of the New Deal agricultural programs. Price supports or cash payments for taking land out of production could help only those farmers who had land to leave idle or who produced some commodity on which the price was being supported. No part of the AAA had anything significant to offer to the hundreds of thousands of sharecroppers, tenants, and those on small acreages in such poor farming areas as Appalachia. In fact, the act contributed to worsening conditions among many poverty-stricken southern sharecroppers. As cotton growers reduced their acreage, they did not need as many sharecroppers, and thousands were forced off the land. By early 1935 one estimate placed the number of evicted sharecroppers and tenants at 100,000. Furthermore, AAA officials did not enforce an equitable distribution of the benefit payments between landowners and sharecroppers. Landless and even homeless, these pitiful people survived on occasional daywork or public relief.[22]

Conditions among southern sharecroppers and tenants served to emphasize the great differences that existed in the economic position of farmers throughout the United States. Among the approximately 6.5 million farm units, a sharp class division existed, a condition recognized by the Census Bureau in 1950 when it began to present statistics on farms by economic class. On the basis of size of farm, production, income, and living standards, a wide gulf existed among different groups of farmers. In 1940 there were 1,519,373 farms of less than 30 acres. This 23 percent of American farmers were mainly farmers in name only. Even the top units in this group harvested an average of about 11 acres per farm and had little or no hope of becoming successful commercial operators.

Another 2,058,337 farms had between 30 and 100 acres each, and their operators harvested from 20 and 36 acres. Most of these farmers were also very poor. Only the largest and most productive in this group were likely to become economically successful, and then only if they received a good deal of outside help from government. Many of the tobacco farms and some specialty farm units which fell in this category earned a small profit, but, overall, these farmers had little hope for economic success. Most of the farms that had less than 100 acres were too small to provide a decent living even if good cropping systems and productivity were maintained.

The nation's poorest farmers, numbering more than half of the total, were located mostly in the South, the Appalachian region, the Great Lakes states, the western Great Plains, and the Pacific Northwest. Many of them farmed hilly, eroded, or cutover submarginal land on which, as the President's Committee on Farm Tenancy reported, families could not make a living "without a considerable amount of public aid." They simply did not have enough production of crops or livestock to earn sufficient income. In one poor cutover area of the Lake states, 63 percent of the farm families received less than $100 per year in the generally prosperous year of 1929. A large number of the poorest farmers were in the South, where they cultivated a few acres of land as sharecroppers and tenants. In perpetual debt to landlords and merchants, and producing only a few bales of cotton or small quantities of other crops, these farmers were bound to permanent poverty. The rungs in the agricultural ladder led downward, not to farm ownership or economic security. Surveying the situation in 1936, President Roosevelt wrote that "we have fallen far short of achieving the traditional American ideal of owner-operated farms." The President did not exaggerate. In 1939 nearly 3 million farm families sold, traded, or used in the household less than $600 worth of agricultural products.[23] In short, about half of the nation's farmers could hardly be considered viable commercial operators.

There were those in the Roosevelt administration who believed that special programs should be devised to help these unfortunate rural residents. As investigations of poverty among farmers proceeded, it soon became clear that the problems of the Great Depression and programs designed for commercial producers had only added to an already bad situation. For years incomes and standards of living had been little above the survival level for millions of farm people. Later known as "economically disadvantaged," they had gone largely unnoticed because they lacked both economic and political power. Help could come either from improving their position within agriculture or through shifting them to nonfarm jobs. The latter was not feasible at a time when millions of urban citizens were unemployed.

To bring hope and economic improvement to at least some of agriculture's "poverty fourth," President Roosevelt established the Resettlement Administration in April 1935. With funds from the Emergency Relief Appropriation Act, the RA set out to help what one observer called "all the dispossessed and disinherited." It retired marginal land, resettled farmers on better acreages, made loans to buy additional land and equipment for a new start, and established model cooperative communities. The main objective of the Resettlement Administration was rural rehabilitation, and by

1936 the agency was touching in one way or another more than 500,000 poor farm families. One distinctive feature of the programs was the degree of supervision which federal workers exerted over the affairs of their client families.[24]

In 1937 the Resettlement Administration was absorbed by the Farm Security Administration set up under the Bankhead-Jones Farm Tenant Act. The FSA took over most of the functions of the RA, but shifted the emphasis somewhat from fighting rural poverty on a broad front to reducing farm tenancy and strengthening the family farm. Liberal credit was provided to help tenants and sharecroppers buy land, and rehabilitation loans were made for purchasing equipment and livestock, and for subsistence when ordinary credit was not available. The FSA also encouraged the formation of consumer and producer cooperatives, and even established some cooperative farming associations.[25]

While the Resettlement Administration and the Farm Security Administration assisted some helpless and low-income farmers to achieve a degree of economic independence, at best these agencies could claim only modest success. The power structure in agriculture—the USDA, the AFBF, the commodity groups, the county agents, the extension service, and the land-grant colleges—never fully accepted either the goals or the means to change conditions among the country's poorest farmers. Consequently, the FSA never had enough money to lift hundreds of thousands of rural residents out of poverty and make them independent, self-supporting farmers. Some southerners in Congress opposed helping poor farmers because it disrupted traditional landlord-tenant relationships. Those who were exploiting black labor did not want to see blacks become independent operators because it would destroy a source of cheap farm labor. Moreover, the FSA became one of the victims of a growing conservative opposition to Roosevelt and the New Deal. In 1943 severe appropriations cuts reduced FSA's activities and many of its projects were soon liquidated. It continued to make some farm loans.[26] But the politics of trying to restore a large element of the poorest farm population had, for the time being, ended. The fight against rural poverty was not seriously resumed until the 1960s.

By the end of the 1930s America's nearly 3 million commercial farmers were the beneficiaries of a vast array of federal programs. These included direct cash payments for taking land out of production, cheap credit, price supports for basic crops and some that were not so basic, surplus food distribution to alleviate surpluses, conservation measures, federally backed marketing orders, aid for rural electrification, support for farmer cooperatives, and others.

The Roosevelt administration gave more direct help to farmers than to any other economic group or class of businessmen. Under pressure from farmers, Congress and the President had injected the federal government into agriculture to an extent thought both undesirable and impossible only a decade earlier. That boasted individualism and independence of farmers had something of a hollow ring by 1940. To a considerable degree landowning producers, and some classes of tenants, had much of the traditional risk in farming considerably reduced by federal programs.

Never before and never again would farmers exert as much political influence in Washington as they did between 1933 and the early 1940s. Depressed conditions in agriculture and a widespread belief that national welfare depended on agricultural prosperity were important elements in the growth of farm political power. However, these conditions and beliefs were not new. The 1930s were different because farmers organized much more effectively than at any time in the past, and took better advantage of every strength they possessed.

A strong President and Secretary of Agriculture who were sympathetic to the problems of farmers and who had widespread popular support provided much of the political muscle needed to get farm programs passed. Moreover, the various measures enacted were developed by specialists in different agencies, not by farmers or farm leaders. During most of the 1930s the American Farm Bureau Federation exerted a major influence in the administration and in Congress. Edward A. O'Neal, an Alabama farmer and president of the AFBF, was a powerful figure in Washington, and his organization gave the USDA needed political clout. While the three major farm organizations were not united on farm policy, the Grange and the Farmers Union joined the AFBF in demanding parity prices. Moreover, influential congressmen and senators from both parties, but especially on the Democratic side, were deeply committed to federal farm programs. Alabama Senator John H. Bankhead, often known as "parity John," successfully pushed a variety of farm bills. "Cotton Ed" Smith, chairman of the Senate Committee on Agriculture, while often critical of the New Deal, was both influential and dedicated to cotton's welfare, and Republican Senator Arthur Capper of Kansas guarded the welfare of wheat growers. In the House of Representatives, Democrats Marvin Jones of Texas and H. P. Fulmer of South Carolina helped look after cotton, peanuts, and other crops, and they received support from such Republican congressmen as Clifford Hope of Kansas, whose interests were in wheat. Add to these sources of power the county agents, the extension service, and the land-grant colleges, and it is clear that commercial farmers had a formidable political voice. However critical some leaders may have been of this rising farm power, they begrudgingly recognized its existence. After trying to reduce appropriations for the purpose of buying agricultural surpluses in 1940, Senator Alva B. Adams of Colorado declared: "the country ought to be very grateful it was only $100 million, because they [proponents] could have gotten $200 million or $300 million just as well. All they would have to do is say it was for the farmers."[27]

While there were differences among those speaking for farmers, because of varying personal philosophies, particular regional loyalties, and different commodity interests, the great majority of agricultural spokesmen agreed on one thing—parity. By the eve of World War II the parity concept approached the acceptance and authority of Scripture in farm thinking. Differences existed as to how parity prices and income might be reached, but there was no dispute about the desirability of the end.

Although farm conditions greatly improved between 1933 and 1940, the New Deal programs did not restore full prosperity to agriculture. Net income from farming to agricultural producers rose from about $2 billion in 1932 to more than double that figure, $4.4 billion, in 1939. Net income per farm jumped from $304 to $685 in the same period. From 1934 to 1940 inclusive, direct federal payments to farmers totaled some $4.5 billion, exclusive of Farm Security-type expenditures. Despite generally improved conditions, farm prices and incomes remained relatively low. By 1939 farm prices averaged only about 80 percent of parity, and the average annual per capita income of people on farms was only 37.5 percent of that received by the nonfarm population. The problem centered around continued heavy production and price-depressing surpluses, despite acreage restrictions. The United States had millions of bales of cotton in storage, enough for a full year of domestic consumption and exports, plus large quantities of other crops. A crisis of abundance in agriculture was at hand by 1940. Secretary Wallace had the idea of an ever-normal granary in which supplies would be built up in lush years to tide the country over in periods of scarcity. The trouble was that there did not seem to be any lean years, and American grain bins and warehouses bulged with surpluses. Some major intervention by government would have been required to save the situation from disaster if the demands of World War II had not wiped out the huge supplies. The onrushing changes in American farming were posing some real threats to the New Deal program of balancing production and consumption, and to achieving fair prices for farmers.

The Quickening Pace of Agricultural Change
The 1920s and 1930s

IV

BEFORE THE 1920S, CHANGE IN AMERICAN agriculture had been slow. To be sure, during the nineteenth century farmers had adopted machines to lighten their burdens, and they used scientific knowledge to improve production of crops and livestock. But the main advances had been spotty and the degree to which the newer agricultural practices had been adopted varied greatly among farms and within regions. For example, mechanization of wheat production had been achieved with the development of horse-drawn seeders, harvesters, and threshers. Some combine-harvesters were also in use. Cotton, on the other hand, still required a great deal of hand labor well into the twentieth century, especially for weeding and picking. Dairying, vegetable production, fruit farming, and many other aspects of agriculture were performed almost entirely by human and animal labor. The machines that were available were simply an extension of manual labor and horse-power. As late as the second and third decades of the twentieth century, the three-mile-an-hour gait of the horse established the speed and power of most field work on American farms.

While there was no typical farm, it is possible to describe the characteristics of a substantial segment of American agriculture by looking at a fairly common type of farm family in Iowa. Assume it was a day in late June, 1925, on the 160-acre farm of John Johnson. Corn and oats, the two main field crops, were growing nicely, cows were grazing in a 40-acre pasture, and pigs were rooting and squealing in a small grassy lot next to the hog barn. Chickens were clucking everywhere around the barnyard. The white frame six-room house was snuggled in a grove of trees, and a red barn housed six to ten horses. To anyone passing by on the dirt road the entire setting gave the impression of stability, security, and comfort. But what was the day like in the lives of the Johnson family?

Johnson and his wife rose at about 5:00 A.M. As Johnson headed for the barn, his wife took a corncob soaked in kerosene, placed it in the cookstove, and started the morning fire on which she would soon cook breakfast. Going into the nearby pasture, Johnson walked through the grass wet with morning dew to get the eight or ten milk cows. At about the same time, he haltered and tied six horses in their stalls. After feeding them oats and hay, Johnson, perhaps with the help of his wife or older son or daughter, milked the cows. He then separated the milk with a hand separator in order to get the cream, which was an important source of cash. Besides, his wife needed cream from which to churn butter. Johnson then carried the several pails of skimmed milk and poured it in a trough for his pigs. While at the hog lot, he threw a few bushels of ear corn from a wagon or granary out to the growing shoats. He returned to the barn and curried and harnessed the horses. Meanwhile, his wife or children had fed and watered the chickens.

By this time it was 7:00 A.M. Farmer Johnson had already put in two hours of hard work, all with his own hands and muscles. After a hearty breakfast, he hitched four horses to a two-row cultivator and headed for the cornfield. The teen-age son or hired man followed him with a two-horse team and a single-row machine. Back and forth across the field they went, guiding the cultivator with their feet to make sure that the shovels did not hit any of the tender young plants. It was hot and humid. Both men and beasts perspired profusely. After several rounds, the men stopped to rest the horses and to take a drink of water from a jug. Then back and forth again until noon.

Before they could eat, the Johnson men had to unhitch the horses and water and feed them. After an hour's break or so, they were back in the field until around 6:00. Arriving back at the farmstead, Johnson cared for the horses. His wife or children had driven the cows into the barnyard for the evening milking. They repeated the morning routine—milking, separating, and feeding the skimmed milk to the hogs. It was now 7:30 or 8:00 P.M. and time for supper.

While Johnson had been in the field, his wife had been busy with her household and other duties. Since she had no running water and it was wash day, she carried several buckets of water from the hand-pumped well outside and poured it into a large tub or boiler on the kitchen stove. She fueled the stove with wood, corncobs, or coal. The fire to heat the wash water made the house uncomfortably hot. She poured the heated water into a washing machine, where the water and clothes were agitated by power from a popping gasoline motor. She was lucky, since most farm women still used hand power and many washed in the yard or on the porch. After washing the clothes, she hung them on a clothesline in the yard, where solar power dried them. Besides washing, which was a Monday job on most farms, she cooked, baked, made butter, canned vegetables, took care of the garden and chickens, customarily considered woman's work, and mended clothes.

By the time the Johnsons sat down to supper at dusk they had worked almost steady for thirteen or fourteen hours. After supper Mrs. Johnson washed the dishes while the children carried in a bucket or two of water for morning. Johnson glanced at the weekly paper that had arrived by mail from the county seat. Reading by kerosene lamp was not easy for tired eyes. By 9:30 or thereabouts the family members made their last trip "out back" because they had no indoor bathroom, and dropped into bed to rest up for another day.

This routine would vary some, depending on whether it was planting, haying, harvesting and threshing, or corn-picking time, but life was characterized more by sameness than by differences. If periods between cultivating and haymaking, or between grain harvesting and corn picking provided some spare time, the milk cows kept by most midwestern farmers governed a farmer's schedule like a tyrant. Only those who have known general farming firsthand can truly understand the restrictive demands placed on farmers by their milk cows. Someone had to be present to do the milking every morning and evening, seven days a week, 365 days a year. "We must get home to milk" was a common rural expression which cut short many social visits and local celebrations.

Amos Jones, a black farmer in Laurens County, Georgia, in the heart of the South's black belt, had little in common with a middle-class Iowa farmer like John Johnson. A sharecropper who cultivated about 30 acres of cotton, Jones and his family faced a constant struggle for survival. Jones and his wife and four children lived in a two-room log house with a leaky roof and holes in the floor. The entire family slept in one room, while the other was used for a kitchen. Household goods consisted of a few pieces of primitive furniture and some cooking utensils. The ground was beaten hard just outside the cabin door, and a few feet beyond was the edge of the cotton field.

When Jones arose on a June morning in 1925, there were no cows to milk. About half of the black sharecroppers in the county did not own a cow. His livestock consisted of a mule, three hogs, and half a dozen chickens. After a breakfast of salt pork, molasses, and cornmeal mush cooked over a fireplace, Jones, his wife, and his two older children stepped outside the cabin, picked up their hoes, and began weeding cotton. After a break for the noon meal, consisting of more salt pork and cornmeal, they worked until late afternoon. For supper they had some vegetables, which gave a little variety to their usual salt pork, cornmeal, and molasses diet. Without livestock there were no farm chores. Every day was hard. It was not that the physical work made life so difficult; there just seemed to be no hope of achieving a better life. Jones represented several hundred thousand very poor black and white farmers in the South who were caught in an economic trap of grinding poverty from which there seemed to be no escape.

The fictitious farm families described here could be found throughout large sections of the Middle West and South. There were very significant variations among farmers, but whether it was a one-mule cotton farm, a dairy in Pennsylvania, a fruit ranch in Washington or Oregon, or a wheat and cattle operation in central Kansas, there was a common denominator: most of the labor was performed by people and animals. Farm labor everywhere was hard and confining. Farming was governed by the seasons, the needs of particular crops, and the demands of livestock. The differences on individual farms, among the crops raised, and within regions were mainly ones of degree.

Despite the hard work and uncertain incomes, life for the commercial family farmer was not without its satisfactions and good times. Many farmers had a spiritual affinity with the land and open spaces. They loved the smell of freshly turned earth in the spring, and the beauty of cotton in bloom or fields of ripening wheat. They saw nature's mysteries unraveled in the birth and growth of livestock and poultry, and they enjoyed neighborliness and community functions at the school or church. Farmers who had economic independence enjoyed being their own boss. They punched no factory time clock and took no orders from an exacting foreman. For some, at least, the Jeffersonian ideal had become reality.

By the 1920s vast and fundamental changes were beginning to emerge on American farms. The application of new technology, chemistry, and plant and animal sciences began to accelerate, producing changes more important than anything that had ever happened in the history of American farming. The combined work of engineers, entomologists, chemists, botanists, geneticists, agronomists, and other scientists completely changed the face of American farming. The evolutionary changes of former years were about to speed up, and in the 1940s and 1950s they became of such fundamental importance as to be called revolutionary. However, most Americans, including farmers themselves, did not fully grasp just what was happening and how the changes would affect farmers and their position in American society.

The most important development in American agriculture in the

1920s and the 1930s was the gradual shift in some parts of the country from horses and mules to tractor power. In 1920 only 3.6 percent of the farmers owned tractors, and very few farmers thought that these chugging, noisy machines would ever replace the faithful horse. In 1925 farmer Johnson in Iowa might have had a tractor, but the chances would have been less than one in ten because only 10 percent of that state's farmers were then using tractors. The introduction of International Harvester's all-purpose Farmall tractor in 1924 was a boon to mechanized agriculture. It was a fairly small, maneuverable machine, suitable for a wide variety of tasks in the field and around the farmstead. In the early 1930s rubber tires replaced steel-cleated wheels and made tractors more efficient and comfortable. By 1930, 13.5 percent of the nation's farmers had tractors; by 1940 it was 23.1 percent. Farms in the Middle West and the Great Plains had the highest percentage of tractors, since corn and small grain crops lent themselves to complete mechanization. More than half of the Illinois, Iowa, Dakota, Nebraska, and Kansas farmers used tractors by 1940, compared to less than 5 percent in North Carolina, Georgia, Alabama, and Mississippi, where cotton, peanuts, tobacco, and other crops resisted mechanization.

Farmers at first hitched their tractors to machinery designed for horses, because there was no specialized equipment available. But this practice reduced the advantages of tractor speed and endurance, as well as the need for less labor. It was uneconomical to pull a cultivator built for horse-power behind a tractor since a person was needed to ride the cultivator and operate it while another drove the tractor. It was not until manufacturers introduced cultivators that could be mounted on the tractor that operators achieved the full advantage of tractor power for that farm work. During the 1920s and 1930s more and more machines and equipment were designed especially for tractors. The development of the power takeoff in the 1920s was of special importance. This device transferred the power from the tractor directly to such machines as harvesters, mowers, grinders, and other equipment. Hydraulic power to lift and lower machinery simply by moving a small lever also improved the efficiency of tractor power. The tractor gradually became something much more than an iron horse pulling the old machinery. It was a source of power attached directly to new and more efficient machines. Tractors made possible the use of larger equipment which could be moved through the field at greater speed completing the process of plowing, cultivating, harvesting, and other tasks better and faster.

Horses on American farms were pushed into second place also by the rapid adoption of trucks and automobiles by farmers in the 1920s. Hauling grain and livestock to market with a team and wagon had always been a slow, time-consuming job. In terms of the farmer's time, transportation costs were high. The use of automobiles and trucks greatly reduced rural isolation and had important social consequences, but from a purely economic viewpoint these machines helped to make many aspects of farm work easier and more efficient. The truck and the grain combine, for example, complemented one another ideally; grain could be delivered to the granary or to market as fast as it was harvested. By 1930 some 13 percent of the farmers had trucks and 58 percent owned automobiles. In the Midwest and the West the percentages were considerably higher.[1] Tractors, trucks, and automobiles introduced a new dynamism into farming. But more than that, they freed men's minds from the concept of the three-mile-an-hour farming associated with horse-power.

Because of relatively low prices for farm commodities during the 1920s and 1930s, there were strong incentives for individual farmers to cut labor costs and to increase their efficiency through the use of new machinery. Those who could reduce the per-unit cost of saleable commodities stood a better chance to make a profit. Besides purchasing tractors, farmers bought combines which harvested and threshed grain in a single operation, better planters, cultivators, discs, drills, mowing machines, and other tractor-powered equipment. For example, in 1938 about 50 percent of the wheat was "combined," compared to less than 5 percent in 1920.[2]

Farm mechanization did not proceed uniformly, but moved ahead by fits and starts. From 1925 to 1930 the number of tractors increased by some 378,000, but only about 50,000 were added by depression-ridden farmers in the years from 1931 to 1935. After 1935 the number of farmers who bought tractors rose rapidly, gaining more than a half million by 1940. In the early 1930s extremely low farm prices, lack of credit and capital, as well as very cheap labor, retarded the rate of farm mechanization. After the New Deal farm programs were well in place and farm income began to rise, mechanization made remarkable advances during the latter half of the decade.

Another major aspect of the budding agricultural revolution during the pre-World War II years was the development of improved strains and breeds of crops and livestock. The best and most important example was the introduction of hybrid corn. Henry A. Wallace was a pioneer in this field. It was a strange contradiction that at the same time Secretary Wallace was cutting acreage to avoid price-depressing surpluses, he was breeding corn that would produce a much more abundant crop. In 1933 it was estimated that only 40,000 acres of hybrid corn were planted in the United States. Within seven years hybrid corn had replaced open-pollinated corn on at least 75 percent of the farms in the Corn Belt. A more vigorous plant, hybrid corn resisted insects and diseases, and initially produced from 15 to 20 percent more per acre. It was clear why corn surpluses were not down significantly in the late 1930s; the increased output per acre made up for the acreage reduction—and even more.

Other advances in crops included the development of Thatcher wheat, which had greater resistance to rust. Improved cotton varieties and higher-yielding forage crops which also had a better protein content were other examples of the new plant breeding. Agricultural scientists also improved the breeds of cattle, swine, and poultry. Although the results came years later, in 1934 crossbreeding experiments were begun to produce a meat-type hog with less fat. By selective breeding and better feeding practices, dairy cow production rose markedly by the end of the 1930s. The average milk output per dairy cow increased in Wisconsin from 5,410 to 5,850 pounds per year between 1925 and 1940. At the same time, mechanical milking machines greatly reduced the man-hours needed in dairying, and refrigeration at the farm site improved the quality of milk.

A third dimension of agricultural change was the increasing use of chemicals in agricultural production. To raise their output per acre, farmers began to apply more lime and commercial fertilizers to their cropland. Although the total was small, nitrogen production rose more than three times between 1921 and 1929 and consumption in the United States nearly doubled during the last five years of the decade. Total commercial fertilizer use from 1924 to 1929 rose by nearly one million tons. During the worst years of the Great Depression farmers used less commercial nutrients, but after 1934 consumption increased rapidly. For example, the use of liming materials rose more than seven times between 1932 and 1940.[3] The New Deal conservation programs contributed significantly to this development. Southern cotton and tobacco farmers had relied heavily on fertilizer for many years, but the most important change was the growing applications by midwestern farmers whose land responded so well to commercial fertilizer treatment. Between 1935 and 1940, for example, Iowa farmers nearly tripled their consumption of commercially mixed fertilizers. The increasing availability and use of insecticides and fungicides helped farmers cut down losses from crop-destroying pests. On the eve of World War II, however, scientists were merely on the threshold of ushering in the chemical revolution on the farm.

Initially, only a few farmers adopted the most advanced agricultural techniques and practices. Time and conditioning were necessary for the spread of technology and new farm practices. In short, there was a gap between knowledge and application. Many farmers, moreover, were not in a position to take advantage of the latest machinery or improved techniques of crop and livestock production. They either lacked the necessary desire and managerial talents, or did not have enough capital or land to make changes practical. An Iowa farmer with 160 or 200 acres of good land could benefit from the purchase of a tractor, while the poor tenant in Georgia with 30 or 40 acres had neither the money or credit for such an investment, nor the labor requirements which would make such a move economically feasible. Although mechanization advanced further than other aspects of the emerging agricultural revolution in the 1920s and 1930s, the combined changes greatly intensified the surplus problem. By the late 1930s better farming practices, combined with good weather, produced tremendous oversupplies. Overall, farm production was up some 10 percent from that of 1929, despite the fact that harvested cropland was *down* by 38 million acres.[4]

Increased farm output, as a result of a wider application of science and technology to agriculture, posed a dangerous threat to the goals of price and income parity established by New Deal planners and Congress. By 1939 the United States had 585 million bushels of corn, 252 million bushels of wheat, and nearly 13 million bales of cotton in storage. Bins and warehouses on and off the farm overflowed. Corn, wheat, and cotton prices were only 59, 50, and 66 percent of parity respectively. Edward O'Neal, president of the American Farm Bureau Federation, told delegates at the organization's 1940 convention that farmers had done everything expected of them, but that parity was still a very elusive goal. Low incomes followed low commodity prices. Per capita cash income from farm marketings averaged only $241 a year in 1939. This was far above the Depression low, but less than in 1936. Despite acreage restrictions, by 1939 and 1940 surplus production and the goal of parity prices or income for farmers seemed to be on a collision course.[5]

As a result of improvements in agricultural productivity, the number of workers in agriculture declined from 25.6 to 17.5 percent of the employed work force in the United States between 1920 and 1940. During the same period, however, total farm population remained about the same, between 31 and 32 million. Despite the movement of millions of people away from the farm, those returning to rural areas during the Depression, and high birth rates, kept the farm population steady. However, as a percentage of the *total* population, those on farms made up only 23.2 percent in 1940 compared to 29.9 percent twenty years earlier. Farm population declined most in the North Central states, where mechanization made rapid advances, and in the West South Central states, where drought and hard times drove many farmers to the West Coast in the 1930s. In the East South Central and South Atlantic states the number of people on farms actually increased in those two decades.

Advances in farm mechanization dramatically brought to national attention the problem of surplus farm population. As farmers adopted the latest machines and enlarged their operations, many farmers and farm workers were displaced. As contemporaries expressed it, they were tractored off the land. This process occurred when a farmer found that he needed more land to get the greatest efficiency out of his machinery. A corn or wheat farmer, for example, finding that he could handle considerably more land with his tractor and other machines, sought to buy or lease additional acreage. The nearby tenant or smaller operator was in no position to compete for land with the bigger farmer with the latest equipment. Consequently, many of the smaller, poorer producers sold out or lost their leases and moved away, leaving the larger farmer in control of the land. With the introduction of rubber tires on tractors and other machines, it was less important that the additional land be contiguous to the farmer's home place. He could move to his newly acquired fields in a few minutes, even if they were several miles away. Retired farmers were often glad to rent their land to a well-equipped neighbor. Consequently, as one Indiana agricultural official said in 1941, "the use of larger tractor equipment has gradually displaced the tenants who used to occupy about every 80 acres." Reports from all over the Corn Belt in 1940 and early 1941 indicated that farms were getting harder and harder to rent. A North Dakota professor explained that when farming no longer could absorb young people in his area, "and after remaining idle parasites on the farmstead for a time, they float into towns and villages, marry, and join the Works Progress Administration forces."[6]

On southern plantations, where an owner might have several hundred acres that had been farmed by sharecroppers in units of 30 or 40 acres, the tendency was to organize larger fields, use tractors and other machines, and leave many of the sharecroppers without land to farm. One study in the 1930s showed that 36 tractors displaced 67 farm families. Some of the former sharecroppers were hired as day laborers during periods of heavy labor requirements, such as cotton picking, but otherwise they eked out an existence from federal relief or miscellaneous odd jobs. Mechanization and acreage restriction together spelled disaster for thousands of sharecroppers. Black farmers were especially hard hit. This whole distressing situation was dramatized in January 1939, when about 330 mostly black displaced sharecropper families camped out along highways 61 and 60 around Sykeston in Missouri's agriculturally rich bootheel. They literally had no place to go.[7]

The economic and social problems associated with the excessive farm population aroused widespread attention and concern in the three or four years before America entered World War II. The more than 6 million farms and 30 million farm people were clearly not needed to produce food and fiber even under the older methods of production, to say nothing of the new conditions surrounding mechanized farming. The Temporary National Economic Committee reported in 1940 that "without counting any further improvements in farm technology, but simply by a more widespread adoption of the best methods now known, it would be possible to release a large number of farm laborers for work in the cities if jobs were available for them." Farmers were becoming a smaller and smaller minority in the total population, but as a writer in *Fortune* magazine stated, agriculture still had "too much capacity, too much product, too much labor."[8] In brief, from an economic standpoint, there was a surplus of farmers. Unlike conditions in the nineteenth century and the first two decades of the twentieth, the depressed and stagnant nonfarm economy could not provide enough jobs for unneeded farm workers.

"The dispossessed are walking the roads of America today," writer Hazel Hendricks observed in October 1940, "refugees, not of war, but of revolution that is turning agriculture into an industry." Some estimates put the "homeless wanderers" as high as 500,000. Hendricks continued that when one viewed the data on unemployed and underemployed persons in agriculture, "it becomes obvious that only a small proportion of the threatened and distressed groups who have previously looked to farming for a livelihood can look forward in the future to security and a decent living in agriculture." Unfortunately, many of them had not had a "decent living" on the land in the past. Writer Hendricks concluded that "most of our disadvantaged rural people must look outside of agriculture for a decent living."[9]

This critical problem received attention not only in the popular press, but at the highest levels of government. Secretary of Agriculture Henry A. Wallace was especially concerned because of charges that the department's acreage restriction programs were contributing to unemployment and destitution in rural areas. While critics did not think the farm to city movement could be stopped, they did blame the Agricultural Adjustment Administration for accelerating unemployment in agriculture. Moreover, some observers accused the USDA and Congress of indifference toward the needs of the poorest farmers, the sharecroppers and tenants. But like others, about all Secretary Wallace could do was to describe the problem. Without new lands to occupy as in the past, or city jobs, he wrote Mrs. Franklin D. Roosevelt, there was a "damning up on the farms of millions of people who normally would have been taken care of elsewhere." Wallace explained that "most of the surplus population can not hope to find place on the land." The ultimate solution, he wrote, would "not be found in making more farms and more farmers, but in making more city employment."[10]

The thought of not having enough opportunities in agriculture for all people who wanted to farm was distressing and threatening to many Americans. What would happen to the hundreds of thousands of people whose roots were in the land? Some farmers themselves felt a twinge of conscience when they expanded at the expense of their neighbors. Asked where farmers without land would end up, a large Illinois operator replied: "I don't know just where they do go. I guess they kind of dwindle off." Another successful farmer said, "of course it's kind of tough on the little fellow, but a man has to look out for himself."[11]

But what about the larger implications? For generations Americans had talked about the importance of a land-owning yeomanry which lent balance and stability to the nation's democratic institutions. Would landless men be radicalized? After a tour in rural areas, the editor of *Wallace's Farmer* reported that he had seen in the eyes of displaced tenants something "that is pretty sure to snap if they are pushed too far." The chairman of the Corn Belt State Tenancy Commission said that if something were not done to help landless people, he was fearful that unrest would "threaten our democracy."[12]

Could the trends in rural America that were so disturbing to some people be halted? Or should they be? In August 1940, an Interbureau Committee in the USDA made a special report on the implications of farm technology. In this study the committee reviewed some of the proposals that had been made to delay technical advances on the farm. These included restrictions on patents, which would regulate the introduction of new agricultural machinery, and a special tax on machines that displaced labor. While these approaches had some popular appeal, they were not considered feasible by the committee. What needed to be done, the committee argued, was to develop "measures of a remedial nature—to seek . . . to reduce the impact and to cushion the effect of these changes upon the disadvantaged groups." The USDA Committee on Technology suggested, among other things, extending the Farm Security Administration programs of farm loans, setting up more cooperative farms, revising the Agricultural Adjustment Act so it would be of more help to small producers, and implementing various rural work-relief programs, especially in land conservation projects.[13]

Authorities in the Department of Agriculture, however, really had little to offer on how to deal with the problem of surplus farm population. Nobody did. The committee report was vague and indecisive. But it would have made little difference what the Interbureau Committee might have suggested. Neither the national political mood in 1940 nor the political power exerted by major farm groups such as the American Farm Bureau Federation would have permitted any massive government attack on rural poverty. To have changed or retarded the course of American farming as the agricultural revolution was gaining momentum on the eve of World War II would have required extensive controls and intervention in areas of land and property holding, federal credit and supervision, and other aspects of agriculture. It would have meant substituting elaborate government planning over advancing agricultural technology and organization, and private choice, that would have been totally unacceptable to a majority of farmers or other Americans. The farmer, as a writer in *Fortune* explained, was becoming "more of an industrialist, and farming more a way to make a living than a mode of life." To attack the tractor, the combine, and other modern farm machinery was like the weavers in England trying to destroy the power loom. Besides, the growing number of individual farmers who were enjoying the benefits of the new science and technology liked the developments, and for better or worse they and their spokesmen held the political power.[14]

What had happened to farmer Johnson in Iowa by 1940? During the previous fifteen years, Johnson had struggled through the Depression, had adopted more modern farming practices in the late 1930s, and was on his way to a fairly comfortable living. By 1940 he had increased the size of his farm from 160 to 240 acres. He had sold all but one team of horses, which he kept more for sentimental reasons than anything else, and was doing his field work on a rubber-tired tractor. Johnson had torn out some of his cross fences to make larger fields and longer rows. The land in pasture and part of that in oats, which he had needed to grow horse feed fifteen years earlier, was now planted to commercial crops or was in a conservation program on which he drew payments from the federal government. His check from Uncle Sam in 1940 was about $1000. The granary was full of hybrid corn, which had made about 52 bushels an acre, more than double what his fields had produced in 1925. It is true that his commercial fertilizer bill was a new operating expense, but increased production more than made up for that outlay. His new mechanical corn picker, one of the few in the area, had eliminated the hard, backbreaking work of harvesting, and a tractor-powered elevator had put corn in the bins so easily that he almost forgot about those years when his back ached after a hard day of scooping by hand. A large portion of his corn was "sealed" under the loan program of the Commodity Credit Corporation at 61 cents a bushel. Farmer Johnson was making more off his corn and hogs, so he had sold most of his milk cows and kept only one or two for the family milk supply. This was part of the specialization that was going on among many midwestern farmers. Moreover, he no longer just threw out ear corn for his pigs, but supplemented their diet of ground grain with purchased feed, rich in nutrition, which increased their rate of growth. When he went to the hog house or barn after a long day in the field, Johnson no longer carried a kerosene lantern to light his way. He just switched on the electric lights. He also used electricity to pump water.

In the house, Mrs. Johnson was enjoying a new way of life provided by electricity. Her eyes were no longer strained in the evening from the yellow glow of smoky kerosene lamps. Every room was bathed in electric light. She had an electric stove, which replaced the old woodburner. Mrs. Johnson no longer had to take the milk and butter to the well or cellar to keep it cool; she had a refrigerator where her dairy and garden produce were kept fresh and tasty. The freezer was full of frozen pork, beef, and chickens, all ready for the roaster or skillet. She had almost forgotten how she used to go out into the yard, throw out a little corn to attract a young chicken, catch it by the leg with a wire hook, wring its neck, dip it in boiling water to loosen the feathers, pluck it, dress it, and finally cook the fowl for supper. On wash day, Mrs. Johnson no longer had to lift and struggle with water and fuel to produce hot water. She had running water, hot from the tap, with which she filled her electric washing machine. Mrs. Johnson also had an electric sewing machine with which she could mend the family clothes and make new garments if she chose to do so. The family could read and listen to the radio in the evening in some comfort. As bedtime approached, there was no need to take the trip to the little house out back, even if it were an improved model built by the WPA; the Johnsons had a bathroom.

It should be emphasized that this scenario of Johnson's farming operations was not typical throughout the United States. The majority of the nation's 6.1 million farmers did not have as much as 200 acres of land, did not have a tractor and a full line of machinery, did not have electricity or modern conveniences in the home. Johnson and his type represented the frontier of agricultural change, which was making such rapid advances among commercial farmers in the West North Central states and among a relatively small group of producers everywhere in the country.

The substitution of capital for labor was producing much of the change on American farms. The expenditures for machinery, fertilizer, electricity, prepared feeds, pesticides, insecticides, and other nonfarm inputs reduced the amount of labor required and increased productivity. Between 1920 and 1940 the labor input in agricultural production dropped from 50 to 41 percent while capital rose from 32 to 41 percent. Land as a factor in production remained about steady at 18 percent. Farmers who had land, and access to capital, along with management abilities, then, were the ones who were taking American agriculture in new directions.[15] The 2.5 million farmers who had little if any of the means of production, those in Appalachia, in the Southeast, and in other sections of the country where pockets of small nonproductive farms existed, had little chance to join the modernization process. They were the farmers left behind.[16]

Farmers in Wartime
Pearl Harbor to Korea

V

THE PROBLEMS OF SURPLUS PRODUCTION, LOW FARM incomes, and excessive farm population that seemed so serious in 1939 and 1940 quickly disappeared under the impact of defense spending and a recovering industrial economy. World War II solved several farm problems at once. In the first place, the need for food and fiber wiped out the price-depressing surpluses. Indeed, the huge quantities of grain which were a burden in 1939 and 1940 became a blessing by 1942 and 1943. Secondly, the call for troops and the need for workers in war plants began to draw down on the excessive farm population. The dam referred to by Secretary Wallace in 1939 which had held many unneeded people in rural areas was at last broken. Some 5 million people left the farms during the five years following January 1940. That was one-sixth of the total 1940 farm population. Many farmers had the machinery and equipment to take on more land as their neighbors headed for war work. Thirdly, wartime demands and good prices created a higher degree of prosperity in much of rural America than farmers had ever known before. At long last they were enjoying that elusive goal of parity.

The impact of World War II on American farmers, however, became clear only in perspective. No one knew just how the outbreak of war in Europe in 1939 and the growing expenditures for defense would affect farmers. When Hitler's troops marched eastward into Poland and later westward into France, surpluses and low prices still plagued American staple crop producers. The main issue before farmers and their spokesmen in 1939 and 1940 was how to protect whatever gains they had made over the last six or seven years, and perhaps expand them. This required continued political activity.

Despite the broad attack on farm problems by the New Deal, commercial farmers, especially those in the Midwest and Great Plains, gradually drifted back to the Republican Party, which they had forsaken during the Depression. Typical of a good many farmers was one overheard cussing President Roosevelt and the New Deal with great emotion during the campaign of 1940. The President and his programs, the complaining midwestern farmer declared, were taking the country down the road to ruin. A neighbor recalled that in 1933 this farmer had been broke and destitute, and that he had been saved by federal credit and price-support legislation. When it was suggested to this New Deal critic that at best he was inconsistent and at worst ungrateful, he denied that anything other than his own hard work and good management had brought about his financial recovery. Variations on this story circulated in many rural communities throughout the United States. The story was often true.

In the presidential elections of 1940 and 1944, President Roosevelt lost much of the farm support which he had drawn in 1932 and 1936. While farmers may have seemed like ingrates to Democratic politicians, they had only forsaken the Democrats, not agricultural politics. Indeed, once commercial farmers had developed organized political power in the 1930s, they guarded their interests in Washington more jealously than ever. There was disagreement among farmers over which party to support, which farm organization to join, or which policy to back, but there was fairly general agreement that farm welfare could not be disassociated from political action.

To be continually effective in Washington, farmers and their representatives recognized that they must become even better organized. As Edward O'Neal told American Farm Bureau Federation delegates at their annual meeting in December 1940, "I hold a deep conviction that agriculture's only chance to survive in this industrial age is through organization." Congressman Clarence Cannon of Missouri told Farm Bureau members that farmers would receive their rights "only by fighting for them." Cannon explained that during 1940 farm leaders in the House had met "with bitter opposition" to parity payments and that victory had come with ever smaller vote margins. He urged farmers to support those congressmen and senators who backed good agricultural programs and nail the hide of those who did not to the political barn door. The new Secretary of Agriculture, Claude R. Wickard, told the same farm audience that not only must farmers be organized, but that they must seek unity within their own ranks. "In the face of the tremendous problems that confront us today," he said, "division and bickering among the farm organizations and the farm groups would be stupid and dangerous." Then Wickard made another important point. He emphasized that farmers must work with other groups because farmers were a "minority" and did not have the power to achieve their objectives alone.[1]

Certainly farm leaders had no intention of permitting the economic dislocations created by the outbreak of war in Europe in 1939 to hurt American farmers. They were particularly sensitive to charges in 1940 and 1941 that rising farm prices were causing the higher cost of living, and they carefully watched actions which would prevent farmers from receiving the full benefits of increased demand. Indeed, farmers looked at the requirements of defense and war as a chance to achieve at long last their most cherished goal—full parity prices. At the same time, in light of burdensome surpluses of corn, wheat, and cotton, farmers were understandably reluctant to increase their production without some guarantees that greater output would not be permitted to depress prices.

As the United States geared up its defense efforts during 1941, the Roosevelt administration nudged farmers toward greater production. Early in the year USDA officials called for more eggs, pork, dairy products, and oil crops. The Department of Agriculture offered to make enough purchases over a six-month period to guarantee base prices. But farmers were befuddled. Just how would a program of expanded production affect them, and how long would demand remain high? These "are the problems that have given agriculture the jitters for the past five weeks," said a writer in the *Farm Journal* in June 1941. As a McLean County, Illinois, farmer observed: "We can easily increase corn acreage 10 percent, boost cattle marketing 10 to 13 percent during the next six months, feed pigs out to heavier weights, step poultry and egg production up 40 percent [and] dairying 10 percent . . . but we want some assurance of full parity prices rather than pegged prices. We want to know whether government policy is going to sidetrack the parity ideal and run off down the same track that wrecked us after the last war." This farmer was referring to the minimum prices set for wheat and cotton in World War I, which in effect became maximum figures. Moreover, farmers feared that prices set by the government to encourage production might somehow replace the main goal, parity. P. O. Wilson, of the National Livestock Marketing Association, declared: "give producers reasonable assurance of parity prices and they will respond with adequate supplies of meat."[2]

At its national convention in December 1940, the AFBF called for prices equaling 100 percent of parity for five basic crops: corn, wheat, cotton, rice, and tobacco. Under this plan, 85 percent of parity would be guaranteed under the price-support program to farmers who restricted production, and the final 15 percent would be paid directly to cooperating producers from USDA appropriations. In May 1941, Senator John Bankhead of Alabama pushed through Congress a modified version of what the Farm Bureau recommended. The Bankhead amendment guaranteed prices for the five basic crops at 85 percent of parity through mandatory government nonrecourse loans. Bankhead called this law "the greatest piece of legislation ever enacted by any Congress of this nation."[3] While this law fell short of the goal of 100 percent of parity, it was an important step toward establishing the principle of fixed or rigid price supports.

Government planners, however, were not pressing for expanded production of basic crops. Indeed, there were huge surpluses of wheat and cotton in 1941. The immediate question centered around how to guarantee prices to producers of vegetables, oil crops, and livestock, who were being urged to increase their output. In July, as Congress considered legislation to extend the life of the Commodity Credit Corporation, Congressman Henry B. Steagall of Alabama and chairman of the House Banking and Currency Committee introduced an amendment that required the Secretary of Agriculture to support the prices of any commodity for which he requested increased production, at 85 percent of parity. This was a further important step by Congress toward providing fixed parity prices for farmers. A writer for *Farm Journal* said that he was "inclined to gasp at the sweeping nature of the legislation and the ease with which it floated through Congress."[4]

By December 1941, farm production and prices were moving rapidly upward. The Commodities Goals Committee, responsible for establishing government production objectives, had set output goals for a number of agricultural commodities for 1942, and recommended increases for such oil crops as soybeans and peanuts (37 and 78 percent respectively).[5] With the government calling for such large increases, it is little wonder that farmers insisted on guaranteed prices. Indeed, the Secretary of Agriculture was authorized to offer price supports above 85 percent on crops for which he wanted to establish strong enough economic incentives to get needed production. Nevertheless, the problem of surpluses was so ingrained in farm thinking, in the USDA, and among agricultural spokesmen in Congress that it was not until 1943 that the United States really achieved something close to full agricultural production.

The needs of defense and war produced sharply higher prices and wages, and brought demands for price controls. Farmers deeply resented the charge that they were responsible for the rising cost of living. They viewed workers who were receiving higher wages and time-and-a-half for overtime as one of the main causes of inflation. There was a growing feeling among farmers in 1940 and 1941 that the administration was much more friendly toward labor than toward farmers. They feared any legislation that would place a ceiling on farm prices while putting a floor under the prices of nonfarm commodities and wages. A writer for the *Farm Journal* said that agricultural spokesmen would settle for no ceiling prices of less than 100 percent of parity so long "as the labor union bosses are not controlled by law."[6]

When Congress began to consider price-control legislation late in 1941, O'Neal and other farm spokesmen argued that price ceilings should not be set on any farm commodity at less than 110 percent of parity. The AFBF president explained to the House Banking and Currency Committee in October that, because of price fluctuations, parity ceilings must be set at 110 percent of parity in order to give farmers an average 100 percent throughout the year. This was simply a way to assure producers full parity. Moreover, O'Neal said that consumers had become so accustomed "to buying food at starvation prices that they had forgotten what constitutes fair prices." It had taken farmers twenty years, he said, to approach their pre–World War I standard, and he intimated that farmers would not sacrifice the parity principle for the benefit of other groups in the economy.

The Emergency Price Control Act of January 1942 included a provision that gave special consideration to agricultural interests and reflected the influence of vigorous farm lobbying. Under the law, the price of no farm product could be set at less than 110 percent of parity, or below the actual price received on three different dates, whichever was higher. This measure gave the impression of providing a special status for agriculture, and farmers came under bitter attack from urban and consumer groups. During the congressional debate, columnist Raymond Clapper wrote that the Secretary of Agriculture could not control farm prices because that "inflames the appetite of one of the most greedy and overbearing of all pressure groups—the farm bloc." Congressman William L. Pfeiffer of New York charged that farm relief had "degenerated into a grand and glorious racket," and he quoted a *New York Times* editorial which accused congressional farm leaders of recklessly exploiting "the national emergency to grab everything possible for agriculture while the getting is good." Responding to such charges, Congressman Clare E. Hoffman of Michigan said all that farmers wanted was "a fair deal."[7]

During the next few months farm prices continued to advance, and by the summer of 1942 widespread demands arose for stricter limits on agricultural commodity prices and food costs. In June a move developed in the House of Representatives to enact legislation that would permit the sale of government-owned and -controlled wheat and corn at less than parity prices as an approach to curbing higher grocery bills. This attempt at what Congressman Clarence Cannon called "a cleverly planned campaign by industrialists and bureaucrats to submarine farm recovery legislation, and to finance war production at the expense of the farmer" was decisively defeated.

Nevertheless, President Roosevelt became convinced that stronger controls must be placed on farm prices, and on September 7 he asked Congress to act. In order to get farmers to accept less than the 110 percent of parity which had been written into law the previous January, Roosevelt's advisors urged him to offer farmers price guarantees during the postwar readjustment. Ever since the calls had been made for greater output in late 1940 and early 1941, farmers had had nightmares of what might happen after the war. The disaster of 1920 and 1921 was fresh in mind. Farmers would respond to the need for food, said a writer for the *Farm Journal* in October 1941, but "thoughtful producers will insist . . . that they not be left in trouble after the boom is over."

To an industry that was in a normal condition of overproduction, it did not seem unreasonable to provide farmers some guarantees in exchange for all-out production during the war. Consequently, the President said that some way should be worked out to "enable us to place a reasonable ceiling or maximum upon farm products but which will enable us also to guarantee to the farmer that he would receive a fair minimum price for his products for one year, or even two years—or whatever period is necessary after the end of the war."[8]

Following the President's speech, farm representatives descended on Washington at the call of the Farm Bureau, the Grange, and the National Council of Farmer Cooperatives, an organization formed in 1929 to represent farmer cooperatives in Washington. Following hearings and floor debate, Congress amended the Price Control Act in October. It placed a ceiling on farm prices, with some exceptions, at a maximum of 100 percent of parity or the highest price between January 1 and September 15, 1942. Farmers and their leaders were generally satisfied with this legislation because many agricultural prices already averaged well above full parity.

Of much greater importance, however, was the amendment that guaranteed farmers price supports at 90 percent of parity for two years after the war ended. As it turned out, this extended the wartime legislation through the crop year of 1948. This Steagall amendment gave farmers the price protection that they so much wanted. As the President had said in September, there must be no recurrence of 1920 and 1921. The significance of this legislation was not properly recognized at the time, even though Steagall called it of "most stupendous importance." The congressional debates were not particularly long or spirited, partly, perhaps, because there was never any chance that this provision would be defeated. Nevertheless, the postwar guarantees written into the price-control law were a further major step in establishing the principle of high, rigid price supports for farmers.

The activities of farm organizations in obtaining this legislation again brought the farm lobby under heavy attack. But agricultural spokesmen did not retreat. "Thank God we have a farm lobby in the United States," said Congressman Steagall. "I hope we keep it." Steagall said that farmers had been "victimized and impoverished" for more than half a century, and it had been only after farmers developed some "organization and some representation here that we have been able to achieve a small measure of comparative justice for farmers." Steagall told delegates at the AFBF annual convention the following December that "the struggle for justice for the farmer has not come to an end. We must never sleep," he said, "until agriculture has secured its proper place in our economic structure and the farmer permitted to enjoy a standard of living suited to his contributions to the national welfare."[9]

There is no doubt but that the wartime agricultural legislation was highly favorable to farmers. The farm lobby was strong and effective. Powerful southern House members such as Steagall, Cannon, Stephen Pace of Georgia, Harold D. Cooley of North Carolina, and Senators Bankhead, Richard B. Russell of Georgia, and Elmer Thomas on the Democratic side never backed down under the charges that farmers were exploiting consumers. Joined by such Republicans as Clifford Hope of Kansas, they insisted that farmers were at last beginning to get what they should have received all along. Except for the Farmers Union, the major farm organizations worked together in a fair degree of unity and harmony in Washington. The recognized agricultural statesman was Ed O'Neal, who received plaudits time and again from congressmen and senators for his work on behalf of farm legislation. James Patton, who became president of the Farmers Union in 1940, was not a part of the inner circle of farm lobbyists. Liberal, idealistic and militant, Patton was out of tune with Farm Bureau and Grange leadership as he expressed concern for poorer farmers and backed such agencies as the Farm Security Administration, which got little support from the other general farm organizations. But, overall, the interests of commercial farmers were in good hands.

With rising prices and greater output, conditions on the nation's commercial farms during the war contrasted sharply with the prewar years. For many farmers this was the first real prosperity they had enjoyed. As early as October 1941, a writer for the *Farm Journal* reported that "many farmers under 45 will have more dollars to spend in October than they have ever had in any single month in their lives." Late in the same year, Secretary Wickard said that the recent upward movements in prices "were like a good rain at the end of a prolonged drought." It may have been regrettable that it took a world war to consume full production from American farms, but such was the case. For many farmers it was a once-in-a-lifetime situation where they could produce at maximum capacity and still receive parity prices. The problems of government regulations, labor and machinery shortages, and other frustrations were submerged under rapidly rising cash returns. Farm prices averaged 105 percent of parity in 1942 and continued upward in subsequent years. In 1946 prices averaged 123 percent of parity. Between 1940 and 1945 net cash income from farming going to farm operators rose from $2.3 billion to $9.2 billion. Annual farm income per worker jumped from $457 to $1,350 in the same period, while the average annual per capita income of all people on farms rose from $245 to $655 during those same years. Even though this was still only 57 percent as much per capita as nonfarm persons earned, it was more than farmers had ever before received.[10] As a result, many farmers bought more land, paid off debts, raised their living standards, and put money in war bonds and other savings.

Roy Snyder, a farmer who lived northeast of Pierre, South Dakota, was an example of what had happened to many farmers during the war. During the 1930s he had lost his farm because of droughts and low prices. In 1936 he did not grow "a damn thing except plenty of thistles." He repurchased his farm in 1940, and by 1946 he owned 1,120 acres and rented an additional 1,920 acres. He sold $35,000 worth of products in 1946. To some extent it was "The Farmer's Time of Milk and Honey," as *Newsweek* reported in October 1946.

While it had taken farmers and the United States Department of Agriculture a little time to adjust from a program of scarcity to one of all-out production, once that turn had been made farmers established output records. Fortunately, they were assisted by good weather. Between 1940 and 1945 the index of gross farm output rose from 108 to 123 (1935–39 = 100). Wheat and corn production were both up substantially, but even larger gains were made in oil crops. For example, production of soybeans increased from 78 million to 192 million bushels. The index of livestock production, excluding horses and mules, jumped from 110 to 139. And this was achieved with less manpower. The output per worker in agriculture during World War II rose 28 percent overall, but in some states it increased more than 50 percent.

The record set by farmers in World War II did not come about because of a large increase in acreage or the development of any new and bold strategies. Crop acreage grew only about 5 percent. But the newer agricultural practices were adopted by a much wider group of commercial farmers. Trends in mechanization continued strong, as the government gave support to farm machine manufacturing. Between 1940 and 1945 more than 500,000 additional tractors moved to the nation's fields, and the number of combines increased from 190,000 to 330,000, and corn pickers from 110,000 to 168,000. Fertilizer production also got priority attention and some farmers more than doubled their applications. Better seed and more pesticides and insecticides also contributed to greater output. Moreover, the federal government paid subsidies on some crops to increase production. In summary, wider application of the knowledge and technology that had accumulated during the 1930s, coupled with wartime price incentives and a degree of patriotism, allowed farmers to provide an abundance for both America and its allies. People in the United States actually had a better diet and a higher level of food consumption during the war than in the 1930s.

While it was only clear in historical perspective, by the end of World War II the basis of traditional American agriculture was being drastically altered. The forces behind the accelerating agricultural changes were leaving little that resembled farming and farm life before the Great Depression. Technology, science, and new farm organization patterns were combining to destroy the old production practices associated with agriculture, and were developing a new farm built more on the industrial model. Agricultural change by the mid-1940s, wrote one perceptive scholar, was becoming "so sharp and decisive" that he called the process of change the "great disjuncture." The explosive rural transformation, he said, was much more significant than the closing of the frontier in the 1890s. Indeed, it may have been a watershed in American history.[11]

But what about the postwar years? How would these modern, progressive, and productive farmers fare in peacetime? Would the old price-depressing surpluses reappear? How could farmers protect their newfound prosperity? These and other questions loomed large in farm country as the legions of Germany and Japan fell before American military might.

As the end of the war approached, American farm and food policy was filled with uncertainties and confusion. One of the main problems facing agricultural planners was determining the extent of postwar demand. Some officials in the Department of Agriculture were fearful as early as 1944 that if all-out production were encouraged large surpluses would soon accumulate. The postwar food needs of American allies, and the extent to which the United States would meet those needs, were unknown. Because of the feeling that surpluses were about to return, one observer explained, the policy was to make certain that "the last GI potato, the last GI pat of butter and the last GI slice of bread was eaten just as the last shot was fired."[12]

By the time the war had actually ended, however, the need for food to ward off starvation among people in the wartorn areas had become abundantly clear. The prospect of famine prompted Herbert Hoover, an experienced food-relief administrator, to exclaim: "It is now 11:59 on the clock of starvation." By the winter of 1945–46, famine stalked Europe. There were also shortages of meat at home. The United States had promised millions of tons of food for overseas relief, but during the spring of 1946 great doubt existed as to whether it could be delivered. Americans were urged to eat less wheat and meat. Observing the growing demand for farm commodities, and being in a good cash position, commercial farmers were slow to sell their wheat, corn, and livestock. To hold their products would probably mean higher prices. In April 1946, the government offered bonuses of 30 cents a bushel to farmers who would market their wheat and corn, but producers did not respond. Finally, to get the needed supplies, the government raised the ceiling prices 30 cents a bushel on wheat and 15 cents on corn. This was what farmers wanted, and within a short time stocks were reaching the shipping terminals, on their way to hungry people in Europe and elsewhere.[13]

In the controversy over continuing price controls into the postwar period, farmers lined up with those who opposed extending price restraints. Because of the tremendous demand for agricultural commodities both at home and abroad, farmers believed they would come out on top even in a general inflationary spiral. Clinton P. Anderson, who followed Wickard as Secretary of Agriculture, came into sharp conflict with the Office of Price Administration, which was struggling to contain the cost of living. Anderson favored abolishing the wartime subsidies which the farmers now heartily disliked. During the war, the government paid some $1.6 billion in subsidies to producers of eighteen commodities. Under this plan, consumers paid a lower price for certain commodities, while farmers received a fair return through the added cash payment. The policy had been highly successful in holding down wartime food prices.

Farmers, however, complained that subsidies left a certain portion of farm income to the whim of government officials, and, furthermore, caused consumers to expect unrealistically low food prices. Congressman Harold Knutson, Republican from Minnesota, was among those charging that the Democrats favored cash subsidies rather than parity prices so they could buy the farm vote. This policy, he said early in the war, permitted "New Deal politicians to dole out public funds for political purposes."[14] In any event, there was strong opposition to continuing the wartime cash subsidies to farmers after the war.

During 1946 a heated political controversy surrounded the question of abolishing farm subsidies and removing price controls. Secretary Anderson, a vigorous and effective fighter for his constituency, argued against continuing subsidies and for loosening price controls over farm products. President Harry S Truman tried to support Chester Bowles, head of the Office of Price Administration, but he fought a losing battle. When the Truman administration refused to abolish controls on meat prices in the spring of 1946, cattlemen engaged in a kind of producers' strike as they waited for higher prices. *Life* magazine editorialized that farmers were trying to destroy price controls and get even bigger returns. Little could be done, according to the writer, because farmers had one of the "smartest and toughest" lobbies in Washington.[15]

During a few weeks in the summer of 1946 when no price controls were in effect, cattlemen rushed their cattle to market as demands from meat-hungry consumers drove prices sky high. When price controls were reinstituted in August, cattlemen again withheld herds from market. President Truman finally succumbed to the public demand for meat, and on October 14 he reluctantly discarded price ceilings on meat. Meanwhile, price ceilings on other farm commodities had been lifted. By the fall of 1946, when commercial farmers went to the polls to vote in the midterm elections, those with land and production were riding on a sea of prosperity.

As the midterm elections of 1946 approached, the Truman administration had alienated most of organized agriculture. Even Farmers Union President James Patton, the most liberal of the farm leaders, abandoned Truman's farm policies in April before the election. When both the House and the Senate went Republican, it was clear that midwestern farmers had at least temporarily deserted the old New Deal coalition.

The repudiation of Truman by many farmers in 1946 did not mean that they and their spokesmen agreed on postwar agricultural policy. Many different approaches to the needs of agriculture were heard loud and clear. The main issue centered around how to deal with surpluses in an industry that had large excess capacity and which was becoming more productive every year. Wartime legislation had settled the issue through the 1948 crop year by guaranteeing the producers of some eighteen crops a minimum price of 90 percent of parity. Moreover, huge requirements for food and fiber in the wartorn nations provided a solid market for several years. But what would happen when supply and demand got out of balance, as they surely would, and surpluses again threatened farm security and living standards? This was the basic question facing farm policy makers in the late 1940s.

The debate that began over farm policy in 1946 continued with only occasional interruption, through the 1960s. Scarcely any question of domestic policy occupied so much time in Congress or befuddled and perplexed so many people in the administrations from Truman to Kennedy as did the farm problem. There was little agreement as to what policies would be best for farmers or for the country as a whole. The agricultural issue was complicated by rapid changes in agriculture itself, by differing and conflicting interests among farmers, and by the basically unfair bargaining position of competitive farmers who operated in an economy that was becoming increasingly monopolistic and noncompetitive. About the only thing that commercial farmers agreed on after the wartime boom was that they were not receiving a fair income for their labor and capital, a view that statistics confirmed. The main difficulty in trying to find a solution to the farm problems after World War II was that there was no answer or policy that would satisfy all interests involved. Since the farm problem was really unsolvable, Congress and the Presidents continued prewar policies that proved little more than stopgap measures, while waiting for some miracle to happen that would expand demand for the products of the world's most productive farm plant.

A number of farm and political leaders, as well as economists, believed that the best long-range approach to helping farmers was to expand consumption. This was the philosophy of abundance. Viewing surplus farm commodities not as a curse but as a national blessing, these spokesmen looked toward using abundant food supplies to feed the undernourished and to improve dietary standards by expanding the school lunch, food stamp, and other welfare programs. Agricultural markets could also be expanded, it was argued, by increasing exports through the Marshall Plan and other foreign aid programs. The idea was to utilize the great resource of food. If Americans could consume more meat and dairy products, some people argued, this would not only provide better diets but reduce grain surpluses, which would be needed to feed poultry and livestock.

On the other hand, many farm leaders and some of the most distinguished agricultural economists believed that rising incomes, full employment, and distribution of food to the poor at home and overseas would not absorb all that farmers could produce. Ed O'Neal and several farm-state congressmen and senators believed that surpluses were bound to return, and looked to restricted production as the only practical way to keep the country from being inundated with surpluses. While the production control advocates believed that distribution of food through relief and social agencies might help some, they had no faith that consumption could provide a lasting answer to surpluses. As one observer declared, "the capacity of the human stomach is no match for the steadily increasing productivity of the American farmer."[16]

The problem of the government maintaining farm prices at 90 percent of parity drew national attention in the embarrassing and costly potato scandal of 1947 and 1948. Because of excessive production of potatoes in 1946, the Commodity Credit Corporation bought 108 million bushels to support the price. The Department of Agriculture disposed of surplus potatoes free or at giveaway prices to school lunch programs, to starch and alcohol plants, and for livestock feed. In 1947 acreage was cut back in hopes of avoiding another year of surpluses. But potato growers planted their rows closer together, applied more fertilizer and insecticides, and harvested another bumper crop. Continued acreage cutbacks were implemented in 1948, but surplus potatoes again plagued farm planners. That year the government purchased 139 million bushels of potatoes to support prices, and before the ordeal ended potatoes had been burned, dyed to keep them off the market, and used for fertilizer. Meanwhile, domestic potato prices remained relatively high. What kind of policy permitted destruction of food when Americans were forced to pay high prices and people around the world were starving? And that was not all. The federal government had spent about $350 million of the taxpayers' money on the potato program between 1946 and 1949. It was, as *U.S. News and World Report* said, an example of the "farm price dilemma."[17]

The unfortunate experience of supporting the price of a perishable commodity such as potatoes pointed up the unwise policy of encouraging production of products for which there was an insufficient market. Indeed, one of the major criticisms of the price-support program was that it tended to fix agricultural production in certain patterns without regard to market demand. To support cotton or wheat prices at levels considerably above the market price, the critics argued, discouraged those farmers from shifting to other crops or livestock. Southern farmers, for example, continued to raise price-supported cotton even when surpluses were large, rather than shift to other crops, grass, and livestock for which there was a better demand.

During 1947, in the midst of the bad publicity over destroying potatoes, USDA officials undertook a careful study of new or modified approaches to farm problems. House and Senate committees also sent teams into the farm regions to seek the views of farmers, economists, bureaucrats, and others on agricultural matters. Growing sentiment emerged for a policy that would continue price supports on basic storable commodities, but at less than 90 percent of parity. Price supports would be set at levels low enough so as not to stimulate production, but at the same time provide a kind of floor under farm prices. Some experts suggested that the floor might be set at around 75 percent of parity. The idea was to let market prices play a much greater role in determining farm production. Farmers should shift their resources to produce for the market, critics of past policies said, rather than for government storage. The crux of this plan was to move away from the high fixed price supports that agricultural interests had first obtained during the war.

In the midst of these discussions, farm political power underwent a major shake-up. When the AFBF met for its annual meeting in December 1947, the organization was split between its southern and midwestern wings. The old alliance between corn and cotton was under severe strain over the question of high versus lower price supports. The conflict stemmed from different economic interests between corn and cotton farmers and illustrates dramatically the problem of fitting a farm program to the desires and needs of different agricultural interests. Cotton producers wanted high price supports and accepted rigid acreage restriction to forestall surpluses. This was an especially favorable policy for large producers in the Mississippi Delta and in other good cotton growing regions where farms were large and efficient. These planters benefitted from high fixed supports and received large cash payments for taking land out of production. Corn farmers, on the other hand, marketed most of their crop through hogs and cattle, and high corn prices offered them little advantage. They could be hurt by low prices, which increased livestock output, but they were more concerned about the diverted acreage in the cotton and wheat country being planted to feed grains which would compete with their own grain and livestock.

Under O'Neal's leadership, and with the support of powerful southern senators and congressmen, the interests of cotton and tobacco growers, who favored high, fixed supports, had been carefully guarded. In 1947, however, O'Neal announced that after fifteen years as president of the AFBF he would not seek reelection. In electing Allan B. Kline, an Iowa corn and hog farmer, to replace O'Neal, the Farm Bureau delegates reflected the growing power of the Midwest. To confirm this shift of power, the convention passed a resolution abandoning high fixed supports and endorsed "variable price supports."[18]

By 1948, as Congress turned to writing new farm legislation, most spokesmen for agriculture agreed that a shift should be made to flexible price supports. This meant that the level of support would be raised or lowered depending on production. Larger crops would mean lower guarantees, and smaller production would mean higher government supports. The flexible-support policy was based on the principle that when farmers saw that higher production would cause price supports to drop they would voluntarily reduce output. Others argued, however, that lower prices encouraged farmers to produce even more in order to have more units to sell at the lower price. That was the only way, it was said, that farmers could obtain enough income to meet their living and operating expenses. But however farmers might respond to flexible supports, most agricultural thinking in 1948 believed that Congress should reject rigid 90 percent of parity and move toward a sliding or flexible price-support program.

As legislation was under consideration in Congress in May and June 1948, it seemed evident that high fixed supports were about to be abandoned. Congressman Clifford Hope of Kansas and Senator George D. Aiken of Vermont introduced the Hope-Aiken bill, which provided price supports at between 60 and 90 percent of parity, or between 72 and 90 percent if marketing quotas were in effect. However, the friends of high supports were just down, not out. Congressman Hope, who was keenly responsive to wheat interests, aided by southerners, was successful in adding to the bill an amendment that postponed implementation of the new flexible supports. Under the Hope provision, prices of basic crops would be supported at 90 percent of parity until January 1, 1950. Only after that date would the flexible supports become law. This meant two more crop years, 1948 and 1949, of wartime farm price supports.

Congress passed the Hope-Aiken bill on the eve of the nominating conventions and the presidential campaign of 1948. Initially, the farm issue did not appear to be attracting major attention from either President Truman or Republican candidate Thomas E. Dewey. By the time the campaign got underway in September, however, farmers had become aroused over falling prices. As early as July 23, *U.S. News and World Report* headed its article on agriculture: "Return of the Big Crop Problem." Record crops, the writer said, "were bringing back the farm problem." A wheat crop second only to the record-breaker of 1947 inundated marketing and transportation facilities. Grain elevators overflowed, railroads could not provide enough cars for shipment, and in some communities wheat had to be piled on the streets. The prospect was also strong for a bumper corn crop. Moreover, tobacco and potatoes were in surplus.

In a situation of accumulating surpluses and declining prices, Truman grasped the agricultural issue and made the most of it. Since storage facilities to hold price-supported crops were inadequate, Truman, in his best give-'em-hell style, blamed the Republicans for that situation. He also charged the GOP with being hostile to soil conservation and rural electrification. According to Truman, the Republican Congress had "stuck a pitchfork in the farmer's back," and destroyed the prosperity that Democrats had provided since 1933. He flayed the Republicans as "gluttons of privilege" who were aligned with Wall Street and who had no feeling or concern for the needs of farmers. As prices weakened, many midwestern farmers who usually voted Republican quietly reassessed their position. While Republican policies were not responsible for large crops and lower prices, farmers had a gut feeling that a vote for Truman and the Democrats might be in their best interest after all. An Iowa farmer told interviewer Samuel Lubell, "I talked about voting for Dewey all summer, but when the time came I just couldn't do it. I remembered the depression and all the good things that had come to me under the Democrats."[19] Truman skillfully aroused the fear of a return to "them Hoover days" in the minds of many grass-roots voters. When the votes were counted, Republicans were stunned. Midwestern farmers had played an important role in Truman's surprising upset. Even though the number of farms and farmers had been declining rather rapidly, the farm vote was still a formidable political force in 1948, as the Republicans learned to their sorrow.

Harry Truman had demonstrated his skill in farm politics. The more important question from the viewpoint of farmers was whether Truman and the Democratic Congress could meet the challenge of farm economics. The superabundance being produced out on the farm under the impact of the developing agricultural revolution had both politicians and economists scratching their heads in search of answers. Agricultural prices were dropping in 1948 and 1949, and under the price-support program the government was accumulating large quantities of unneeded wheat, cotton, dairy products, and other commodities at great cost to taxpayers. No one really seemed to know what to do. Most leaders believed that farmers needed some price protection, but how could this be achieved without encouraging excessive output, and increasing program costs. During debate over the Hope-Aiken bill, Senator Aiken said that to continue 90 percent of parity, with all of the problems that the policy created, "the time will not be far distant when the American people will rise up and say they will no longer have any farm price support program."[20] The truth of that prediction, of course, would depend on the effectiveness of farmer political power.

Although some farmers and their spokesmen were having second thoughts about flexible price supports even before the presidential election, a return to fixed supports at 90 percent of parity had not been an important issue in the campaign. As prices declined during the fall of 1948, however, more and more voices called for a return to fixed supports at 90 or in some cases 100 percent of parity. Between November 1948, and the spring of 1949, James Patton of the Farmers Union became the main spokesman for 100 percent of parity. Meanwhile, Patton's good friend Charles F. Brannan was appointed Secretary of Agriculture and was carefully studying the whole range of farm problems. Brannan, a liberal Denver lawyer who had moved up through the agricultural bureaucracy, presented his plan to Congress and the nation in April 1949.

Secretary Brannan suggested that the prices of basic commodities be supported at 90 to 100 percent of parity, but that farmers receive supports on only their first $25,700 worth of products. This provision was to insure that the smaller, family-type farmers received most of the benefit. Secondly, Brannan proposed to extend price supports on perishable commodities. Unlike the requirement for basic crops, however, there would be no production controls on such things as livestock, fruits, and vegetables. Farmers could produce as much as they pleased and prices would be determined by the market. In order to guarantee producers parity prices, the government would make a direct payment to farmers amounting to the difference between the market and the parity price. If eggs, for instance, brought

15 cents a dozen on the market and the parity price was 20 cents, the government would pay each egg producer 5 cents a dozen. This idea was not unlike what had been done during World War II, but in that case the purpose of the payment had been to stimulate production.

Brannan argued that his plan would provide good incomes to farmers and at the same time lower prices to consumers. Poor people, especially, would benefit from the production of abundant, cheap, and highly nutritious food, he said. Moreover, Brannan reasoned that increased demand for perishables would cause farmers to shift out of grain and cotton, and produce more of those commodities, such as livestock and dairy products, fruits and vegetables, for which there was a stronger demand. As one student observed, the Brannan plan "was an ingenious effort to wed the concept of abundance to the demands of the Administration's farmer constituency—an attempt to give consumers more food and farmers higher income."[21]

The Brannan plan for farm relief sent shock waves throughout the country. No farm bill since the McNary-Haugen days in the 1920s aroused so much public controversy. It elevated farm policy to one of the most important questions of domestic politics, and created deep divisions and antagonisms within agriculture. Of all the problems facing the United States, said a writer in *Fortune* in June 1949, "the agricultural problem is probably the most vexing." While Brannan received strong backing from the Farmers Union, organized labor, and some northern urban Democrats who were attracted by the prospect of cheaper food, most of the political power structure both within and outside of agriculture rallied against the proposal.

Critics charged that the plan would cost billions, that it would inject government excessively into agriculture, and that it would place farmers in the position of receiving direct federal handouts. Conservative southern democrats and midwestern Republicans fought the plan partly because it was being pushed by a liberal-labor coalition which was supporting a wide variety of social legislation anathema to conservatives. Large cotton and wheat farmers criticized the Brannan approach because of the limit on the amount of government payments that producers could receive.

However, probably more than anything else, opposition from the American Farm Bureau Federation spelled doom for the Brannan plan. The Farm Bureau had opposed direct subsidies to farmers for years, and by 1949 AFBF President Allan Kline expressed the view that there should be less, not more, government in agriculture. Also by that time the Farm Bureau was fully committed to flexible price supports as the basic farm policy. Furthermore, Farm Bureau officials had played no part in writing the Brannan plan, which had been developed by Brannan, his associates in the USDA, and James Patton. Not only did Farm Bureau officials resent the presumption that Department of Agriculture bureaucrats would tell Congress what farmers needed or wanted, they were offended by the close relationship between Brannan and Patton. By 1949 the Farm Bureau and the Farmers Union disagreed on about every major public policy. While the Farm Bureau spoke for the larger, more prosperous commercial farmers, the Farmers Union concentrated its attention on doing something for the smaller and poorer farmers, and also backed liberal social and labor legislation opposed by the Farm Bureau. In any event, the AFBF mustered its tremendous resources against the Brannan plan and did much to kill it. The Grange and other farm organizations also added their influence to the scheme's demise.[22]

Although the House defeated the Brannan plan in the summer of 1949, Congress did not want to adjourn without offering something to their farm constituents. During discussion over a new farm bill in the early fall, there were numerous demands both in Congress and outside Washington for 90 or 100 percent of parity prices at least on six basic commodities—wheat, corn, cotton, rice, tobacco, and peanuts. Senators Russell of Georgia and Milton R. Young of North Dakota, representing cotton and wheat respectively, pushed an amendment through the Senate on October 4 which would have scuttled flexible supports and pegged the price of basic crops at 90 percent of parity. This narrow victory by the advocates of high, fixed supports, however, was later reversed. The Agricultural Act of 1949 continued the principle of sliding price supports but extended the 90 percent of parity prices on basic crops for an additional year, through 1950. The Secretary of Agriculture was authorized to support the prices of some nonbasic commodities at between 60 and 90 percent of parity, and others at between 0 and 90 percent. Acreage allotments were retained and marketing quotas could be implemented if two-thirds of the producers of one of the six basic crops voted favorably in a referendum. Despite all of the talk about revision, the principles legislated in the 1930s still governed farm policy in 1950.

Congress had really failed to come to grips with the farm problem. Under a multitude of political and economic pressures, lawmakers seemed incapable of finding solutions to problems of superabundance. Congress responded to constituent pressures and worked out compromises that kept the more influential commercial farmers fairly well satisfied without running up program costs to the point where nonfarm power blocs would endeavor to overthrow the whole price-support structure. It was policy making by bargain and compromise. From the viewpoint of the nation's approximately 2 million better-off commercial farmers, the result was reasonably satisfactory. Leaders inside and outside of Congress paid lip service to doing something for small, poor operators, but little was done beyond making studies and delivering sympathetic speeches.

Price-depressing surpluses caused farm planners sleepless nights in 1949 and early 1950. However, war again came to the rescue of farmers as it had a decade earlier. With the outbreak of the Korean War in the summer of 1950, demands for most farm commodities rose sharply and gave policy makers a breathing spell. The heavy carry-over of 6.8 million bales of cotton from 1949 was cut by a poor crop in 1950 and the added wartime demand. Wheat and corn stocks were also lowered. As farm prices rose—cattle, cotton, hogs, and rice were all above parity in 1951—the Commodity Credit Corporation was able to dispose of its supplies with little if any loss. Farmers, especially cattlemen, worked vigorously against the administration's price-stabilization efforts. On July 15, 1951, during the controversy over price controls, the *New York Times* reported that "beyond question the single greatest force on Capitol Hill now is the farm bloc." While that assessment was an exaggeration, agricultural power was evident as farmers fought to maintain their little economic boom. After dropping to only $13.6 billion in 1949, net farm income rose to $14.7 billion in 1951 and was nearly as high the following year.[23]

With the pressures of surpluses temporarily removed, Congress again postponed implementing the flexible price-support law. In July 1952, lawmakers amended the Agricultural Act of 1949 to require the maintenance of prices at 90 percent of parity on basic commodities through the crop year of 1954. Moreover, this legislation provided that the highest parity figure again be applied on the six basic commodities—wheat, cotton, corn, rice, tobacco, and peanuts. In 1948 Congress provided for a new calculation of parity which would include prices for the previous ten years as well as for the 1910–14 base period. The so-called new parity meant higher prices for tobacco and rice but lower prices for corn, cotton, wheat, and peanuts. Congress was equal to the task of not offending any of the commodity groups and simply required that through December 31, 1955, prices on basic crops should be set at the higher figure computed under either the new or the old parity formula. There was no indication of diminished farm power in Washington in 1952.

The agricultural issue did not rate high in the Eisenhower/Stevenson presidential race of 1952. Commerical farmers were enjoying relatively good times, and other questions attracted voters' attention. Both party platforms flattered farmers and their importance to national welfare. The Democrats pledged a return to price supports at not less than 90 percent of parity on basic crops, while the Republicans promised to seek full parity in the marketplace. Eisenhower cut the political ground out from under the Democrats on September 6 when he told 100,000 farmers at the National Plowing Contest in Minnesota that he and his party stood fully behind the law to continue 90 percent parity prices through 1954. Explaining his position more fully a little later, Eisenhower said that he favored prices at 100 percent of parity, but that this goal should be achieved in the marketplace. The farm question was overshadowed in 1952, however, by Republican charges against the Democrats of crime, corruption, and communism. Farmers, except in the traditionally Democratic South, voted heavily for Eisenhower even though they were not clear on just how he might assure prosperity on the farm.

Trends in American Farming, 1935-1980

Year	*Number of farms (thousands)*	*Farm population (thousands)*	*Farm population as percent of total population*	*Net income of farm operators from farming (millions of dollars)*	*Per capita disposable income from all sources of farm population (dollars)*	*Per capita disposable income of farmers as percentage of that of nonfarmers*
1935	6,814	32,161	25.3	5,278	237	44.5
1940	6,350	30,547	23.1	4,482	245	36.7
1945	5,967	24,420	17.5	12,312	655	56.9
1950	5,648	23,048	15.2	13,648	840	58.1
1953	4,984	19,874	12.5	12,980	914	54.9
1956	4,514	18,712	11.1	11,254	877	47.7
1960	3,963	15,635	8.7	11,518	1,083	53.8
1963	3,572	13,367	7.1	11,770	1,364	62.4
1965	3,356	12,363	6.4	12,899	1,692	68.2
1966	3,257	11,595	5.9	13,960	1,894	71.7
1967	3,162	10,875	5.5	12,339	1,925	69.0
1968	3,071	10,454	5.2	12,322	2,099	70.5
1969	3,000	10,307	5.1	14,293	2,332	74.0
1970	2,949	9,712	4.7	14,151	2,520	74.1
1971	2,902	9,425	4.6	14,633	2,722	75.0
1972	2,860	9,610	4.6	18,665	3,244	83.9
1973	2,823	9,472	4.5	33,349	4,700	110.2
1974	2,795	9,264	4.4	26,130	4,355	93.5
1975	2,521	8,864	4.2	24,475	4,520	88.4
1976	2,497	8,253	3.8	18,682	4,314	77.7
1977	2,456	7,806	3.6	17,829	5,262	87.1
1978	2,436	6,501	3.0	26,081	6,422	96.2
1979	2,430	6,241	2.8	30,959	7,535	102.4
1980	2,428	6,051	2.7	22,000*	**	**

From USDA, Economics, Statistics, and Cooperatives Service, *Farm Income Statistics*, Statistical Bulletin No. 627 (October, 1979). Some definitions of farms and farm population were changed in 1977, so the figures before and after 1977 are not exactly comparable. See USDA, Economics and Statistics Service, *Economic Indicators of the Farm Sector*, Income and Balance Sheet Statistics Washington, December 1980), and Production and Efficiency Statistics (Washington, February 1981).
*Preliminary
**Not available

Problems, Progress, and Policies in the 1950s

VI

THE KOREAN WAR HAD OFFERED ANOTHER respite to agricultural policy makers, but by the time Dwight D. Eisenhower took office the farm issue had reappeared. By early 1953 average farm prices were about 10 percent lower than they had been in early 1951, and cattle prices were about 30 percent under those of a year earlier. On December 12, 1952, more than a month before Eisenhower's inauguration, *U.S. News and World Report* said that "trouble is brewing for Eisenhower down on the farm." Farm problems were "the troublemaker for American presidents," continued the writer. A few weeks later, *Time* judged that Eisenhower had two major problems—Korea and declining farm prices.

Some people may have agreed with Congressman Usher L. Burdick of North Dakota that Eisenhower knew less about government than "I know about the hereafter," but it soon became evident that the President and Ezra Taft Benson, his new Secretary of Agriculture, had very firm views on the perplexing farm problem. With Eisenhower's full support, Benson indicated that the administration planned to move away from high, fixed supports as soon as possible. With many other authorities, Benson insisted that the 90 percent of parity policy was not economically sound. The main weakness, Benson argued, was that it encouraged an uneconomic allocation of resources. Farmers produced huge quantities of unneeded commodities supported by the government, rather than shifting land, labor, and capital to other products for which there might be a better market. Large government costs, high consumer prices that cut consumption, and reduced exports were other results of high, fixed price supports. Finally, Benson talked about giving farmers freedom to farm, freeing them from quotas and restrictions, and concentrating production not "for government bounty but for a free market." The entire thrust of Benson's argument was that market prices should govern returns to farmers and that price supports should serve mainly as "insurance against disaster." This kind of farm program fit into the basic Eisenhower-Benson philosophy of reduced government expenditures and getting the government out of business.

Many farm economists, agricultural leaders, and othes had been advocating some or all of these same principles since the end of World War II. No one, however, had proposed to change the course of farm policy with more force or conviction than had Benson. Born on an Idaho farm and rising to leadership in agricultural cooperative work, this apostle of the Mormon Church believed deeply in the efficiency of American farmers and in their ability to handle their own affairs. He strongly rejected strict government controls. But the Secretary's campaign to achieve flexible supports came at an inopportune time. In leading the effort to reduce the role of government in agriculture, Eisenhower and Benson were demanding a sharp break with the extensive government planning inaugurated by the New Deal and continued by the Fair Deal just when prices were declining and farmers found themselves in a harsh cost-price squeeze. In 1953, tens of thousands of commercial farmers wanted stronger assurances that prices would be supported at profitable levels, not Benson rhetoric about "the principles, benefits and values of private competitive enterprise."[1]

Benson's talk about returning to a free market set off what the *Saturday Evening Post* called a "nationwide offensive" against the Secretary, who was "catching plain unshirted hell." Indeed, there were widespread calls for Benson's resignation. The Democrats, supported by James Patton and the Farmers Union, had a field day attacking Benson. Some midwestern Republicans were no kinder to the Secretary. Iowa Governor William S. Beardsley said the only thing flexible or sliding supports would be good for "is to flex and slide Benson out of office." But with Eisenhower's unshaken support and with backing from the American Farm Bureau Federation and the Grange, Benson stood firm. During 1953 he went ahead with studies and plans to implement flexible supports.

Farmers, meanwhile, became increasingly angry as agricultural prices slipped downward. In October 1953, a cattle caravan sponsored by the Farmers Union marched on Washington demanding 90 percent of parity prices for cattle. While the principal livestock associations did not want price supports, the Farmers Union was able to get representatives from thirty-two states to descend on the nation's capital and confront Benson. The Secretary rejected their demands, but said he would study the situation. Cattlemen said they had received the "old runaround." Before the ranchers returned home vowing to build a "grass-roots fire" under Benson, they heard a rousing speech by Senator Robert S. Kerr of Oklahoma. Kerr told the cattlemen that Eisenhower did not know much about agriculture when he took office, and that after being advised by Benson he had barely held his own.[2] Kerr said he was for 100 percent of parity.

An even more startling example of farm discontent occurred in Wisconsin's ninth congressional district. At a special election held in October, voters in this heavily rural district elected a Democrat for the first time in the district's history. According to one observer no one took Democrat Lester R. Johnson seriously until he touched the discontent over Benson and farm policy. Johnson's victory was interpreted as a clear rejection of the Benson policies and an endorsement of high, fixed price supports. For the Republicans, columnist Thomas L. Stokes wrote, the election was like a "political atomic blast that mushroomed up over Wisconsin."[3]

By late 1953 and early 1954 the debate over flexible versus rigid price supports had ballooned into a bitter and highly controversial issue. The national press was filled with discussions of "the farm problem." In the autumn of 1953 members of the House Agriculture Committee toured agricultural areas and claimed that they found farmers strongly against flexible supports. Divisions over policy erupted within the farm organizations. For instance, state farm bureaus in Kentucky, Tennessee, and North Dakota called for 90 percent of parity in the face of strong backing for flexible supports by the national organization. Governor Sigurd Anderson of South Dakota, a strong Republican state, announced that he favored 90 percent of parity. Early in 1954 Anderson said that if South Dakota farmers could vote on the issue, they would vote overwhelmingly to continue the 90 percent price-support law. At a meeting of farmers in Dickinson County, Kansas, in November 1953, 114 favored 90 percent of parity, 15 favored flexible price supports, and 27 wanted some kind of two-price system.[4]

It was in this atmosphere that President Eisenhower outlined the directions of his farm policy. In his State of the Union address on January 7, 1954, and four days later in a special message, he left no doubt about his views. He said that the choice before Congress and the country was to curb surpluses by enforcing more stringent acreage reductions and placing rigid quotas over farmers, or by letting the market play a greater role in production planning by farmers. He recommended flexible price supports, a national food reserve, and increased consumption through school lunch, disaster relief, and other programs. "I have chosen this farm program because it will build markets, protect the consumers' food supply, and move food into consumption instead of into storage," he concluded. On January

11, Eisenhower announced that the 90 percent of parity supports would be permitted to expire at the end of 1954, and that the Secretary of Agriculture would adjust price supports to a sliding scale of between 75 and 90 percent. Whether the Benson program would be of any help could be debated, but one thing was certain: the country was being swamped in a sea of surpluses. By early 1954 the government held about 1 billion pounds of dairy products, 800 million bushels of wheat, 900 million bushels of corn, and 9.6 million bales of cotton.[5]

Those who favored continuing 90 percent of parity immediately denounced the President's program and predicted that it would not pass. Senator James O. Eastland of Mississippi said it was "dead as a doornail," and Congressman Carl Albert of Oklahoma exclaimed that it "won't get anywhere." Georgia Senator Richard Russell declared that the plan would not work. Some midwestern Republicans agreed. Senator Milton Young of North Dakota said that if the President's legislation got on the House floor and someone offered a 90 percent of parity amendment, "you couldn't stop it." While a good many Democrats agreed with Eisenhower's economics, politically very few of them could afford to back him. As columnist Roscoe Drummond wrote, several Benson critics believed that he was "economically sound" but "the dominant opinion of farm bloc congressmen" was that flexible supports were "politically poisonous."[6] It appeared as though by opposing the Benson-Eisenhower program, the Democrats had a winning issue for the fall midterm elections.

As Congress considered some revision of farm policy, almost every aspect of the agricultural issue received wide publicity. Popular news magazines carried stories on the volume of surpluses, the cost to taxpayers, which included about $800,000 a day in storage charges, and how government programs were mainly helping the large farmers and corporate operators. *U.S. News and World Report* on May 21 referred to the farm situation as "A 6-Billion Dollar Headache," while other widely read magazines told of storing surplus crops in ships, airplane hangers, and anywhere else that space could be found. After describing the volume of surpluses, a writer in *Look* declared that "unless Congress works its way back to common sense, the potato scandal will be as nothing compared to the outcry over the stifling abundance of wheat, corn, cotton, and dairy products." He concluded that to maintain high price supports when the nation had more food than it could consume was "economic lunacy."[7]

Partly to discredit fixed support prices, early in the summer of 1954 Secretary Benson publicized how a few farm producers were receiving most of the government payments. The records showed that in 1953 there were 64 large corn, wheat, and cotton operators who received more than $100,000 each in government loans for a total of nearly $16 million. The largest payment to any grower was a $1,269,492 loan on 7,220 bales of cotton to the Delta and Pine Land Corporation of Mississippi. In contrast, the average cotton grower in Mississippi received only $372 in loan payments. The Thomas D. Campbell wheat farms in Montana got a loan of $348,646 against 184,516 bushels of wheat, and the largest corn loan in Iowa amounted to $190,944. The average corn payment in Iowa was $2,154.[8] The message was clear. It was the "factories in the fields," those least in need, who were getting most of the benefits from price supports.

Congress discussed the Eisenhower program from March to August 1954. After some bitter debate and several compromises the lawmakers finally approved flexible price supports on the basic crops. The range was to be 82.5 to 90 percent in 1955 and 75 to 90 percent thereafter, except for tobacco, which was to be supported at 90 percent of parity. Supports were also mandatory on some nonbasics at 75 to 90 percent, and other commodities could be supported at the discretion of the Secretary of Agriculture at between 0 and 90 percent. In order to keep current surpluses from being used in computing parity, $2.5 billion worth of those commodities were to be placed in a set-aside. Transition to the new parity would start January 1, 1956. Since this would mean lower support prices for some basic crops, the law prohibited declines of more than 5 percent in any one year.

During the debate, the farm lobbies were much in evidence. The AFBF and the Grange supported the President's program, while the Farmers Union plugged for 90 to 100 percent of parity. Lesser groups were also active. The administration wanted to end mandatory price supports on honey and tung nuts, but the interested producers were able to make enough political trade-offs to keep them in the bill. As every President had discovered since 1933, it was not easy to reconcile agricultural economics with farm politics.

In separate legislation and with little controversy, Congress passed PL 480, which provided for $1 billion to dispose of surpluses overseas in exchange for so-called soft or nonconvertible currencies. The Farm Bureau strongly supported this law as a means of reducing surpluses.

No careful observer believed that the shift from rigid to flexible price supports would improve the farm situation any time soon, if at all. Walter Lippmann wrote that about the only difference was whether farm income would "be subsidized at the congressional rate or at the administration rate." Under the impact of huge surpluses, farm prices continued downward in 1955. By fall, hog prices had dropped to a nine-year low, and in October the government stepped in and bought pork and lard to halt further declines. Total net cash income to farmers from farming dropped from $9.4 billion in 1951 to $7.5 billion in 1955. Disposable personal income of farm people from all sources averaged $988 in 1951, but only $848 four years later. By 1955 average per capita income of farmers was only 48.2 percent of that enjoyed by the nonfarm population. In 1951 it had been 64.4 percent. It is not surprising that farmers were heard to say that "things were going to pot." By late 1955 Secretary Benson again came under withering attacks, and both Democrats and Republicans demanded his resignation. Many of these critics were calling for a return to high, fixed price supports, a move that one observer said would be like "trying to cure a drunk with another drink."

Millions of farmers were in a vicelike cost-price squeeze. Between 1951 and 1956 agricultural prices dropped about 23 percent while the prices of nonfarm commodities stayed about the same. As one dairy farmer said, "I wouldn't mind taking a cut [in milk prices] if I could get a cut in the price of a tractor." But that was not the case. Researchers at Iowa State University studied 140 Iowa farmers and found that the average net income had slid from $10,247 in 1953 to $7,051 in 1955. In parts of Illinois the story was even worse, and in Kansas surveys showed that farmers' net income had dropped 62 percent in the same period. The farmers' inability to control production to the market, to set their prices, or to bargain effectively with nonfarm businesses placed them in an unfavorable position that even government programs had not corrected. What made so many farmers angry at Benson, as Samuel Lubell wrote after visiting many producers, was "his failure to display any sympathy for these special handicaps that farmers labor under."[9]

Farmers complained that they actually subsidized the rest of the economy by providing cheap food and raw materials. Low farm prices had kept food costs low and contributed to the stable price level from 1953 to 1956. Some economists and government leaders admitted this situation. "The great shame to me" said McChestney Martin, chairman of the Federal Reserve Board, "was that we kept stability in the dollar from 1953 to the early part of 1956 by a decline in farm prices, which was being offset by a rise in manufactured prices." This condition seemed unfair to agricultural producers. Farmers also reacted angrily when they were blamed for high food costs. Agricultural spokesmen pointed out that farmers received only about 40 cents out of each food dollar, and that even if some farm commodities were given away it would do very little to reduce grocery costs. For example, farmers pointed to the fact that even if wheat were given to millers and bakers, bread would only be 3 cents a loaf cheaper. The 30 cents worth of cotton in a $3.95 shirt, they said, was not what made clothing cost so much. Farmers believed that they took criticism from consumers that rightfully should have been directed toward labor, transportation, processors, wholesalers, and retailers.[10]

The growing charges that farmers were getting fat on the public dole aroused even sharper responses from the grass roots. Articles such as John Fischer's "The Country Slickers Take Us Again," which appeared in the December 1955 *Harper's*, made farmers and their friends downright mad. Fischer said that politicians in both parties would soon be "groveling all over the barnyard as they count the country vote." Farmers had become accustomed to getting their "dole," Fischer continued, and they acted as though it was theirs as a matter of right. "When any hog keeps his jowls in the trough long enough," he wrote, "he gets to thinking that he owns the trough." Farmers responded that their subsidies were meager compared to those showered on the railroads, airline companies, the post office, newspapers and magazines, and other businesses, including tax breaks for some large corporations.

With an approaching presidential election and cries for help coming from the grass roots, it is not surprising that President Eisenhower proposed new legislation early in 1956. Moreover, it appeared that something must be done to control the accumulating surpluses. Between October 1954 and November 1955, Commodity Credit Corporation holdings rose from about $4 billion to $6 billion. Somehow farmers had to be encouraged to take more land out of production. In Janaury 1956, Eisenhower presented his soil-bank plan. While still insisting that farmers should move toward a market less dependent on government, the President recommended paying farmers for taking acres out of production. This would cut surpluses, put cash in farmers' pockets, and protect their incomes against low prices and crop failures.

Once legislation had been introduced, administration forces could not keep Congress from enacting legislation calling for a return to 90 percent of parity. Despite the fact that the presidential election was only six months away, and news writers reported "widespread farm unrest" in the normally Republican Midwest, Eisenhower promptly vetoed the farm bill. He admitted that his program of flexible supports was no panacea because none existed, but he declared that a return to high, fixed supports would only add to the surplus problem. He urged Congress to pass a soil-bank measure without the unacceptable parity provision as "promptly as possible." Recognizing the futility of any other course, Congress enacted the soil-bank bill in May.

Under this law, producers of basic crops—cotton, corn, wheat, peanuts, rice, and tobacco—could put land normally planted to those crops in an acreage reserve which could be used for no other crop. Farmers received a fixed cash payment for retiring this land. In 1956, for example, payments were $40.05 an acre for corn land, $45 for cotton land, and $19.80 for wheat land. A second part of the law permitted farmers to put entire farms in a conservation reserve for a period of between three and fifteen years. (As it turned out, the maximum contract was for ten years.) This program was to run until 1960. Another section of the bill placed a minimum below which cotton acreage could not be reduced. This was designed to keep small farmers from losing any more acreage under their cotton allotment. As Senator Russell and others explained, acreage reductions had become so serious that many small farmers did not have enough land left to produce a quantity sufficient to pay operating and living expenses.

Some agricultural officials hoped that the soil bank would encourage small farmers to take their entire farms out of production. This, it was believed, would result in lower output, would strengthen prices, and would reduce government storage expenses. Moreover, if some of the smaller farmers rented their land to the government they might leave farming altogether. Secretary Benson had stated frankly that he believed many marginal farmers should seek employment outside of agriculture.

The prospect of farmers renting their land to the government and moving away aroused much of the opposition directed against the soil-bank program. Tenants, it was said, were in special danger of being forced off the land as owners rented to Uncle Sam for a sure cash return. Critics presented the soil bank as being a major factor in reducing population and weakening rural communities. First some of the farmers would leave, then small businesses supplying farmers would close or go broke and vacant store fronts would give towns a ghostlike appearance. In the minds of many people, the soil bank would contribute to both economic and social problems in rural America. With a declining farm population, rural communities were undergoing drastic transformation under the normal course of events. These developments, critics said, should not be encouraged by a government policy which seemed to weaken the family farm, destroy the small town, and threaten fundamental American values. Despite these fears and complaints, farmers eagerly signed up to get their share of government cash. At some county offices, farmers were in line by 5:00 A.M. to be sure their applications were filed before the money ran out. In Eisenhower's mind the main purpose of the soil bank was to cut surpluses, but many viewed it as simply another way to increase the cash flow to farmers. One agricultural official remarked in June 1956, that "you'd be amazed at the terrific political pressures bearing down on us just to give money away."[11]

The soil bank did not provide much help for farmers in 1956, and the Democrats believed that the administration's rejection of 90 percent of parity would give them a good issue in the presidential election. Senator Aiken of Vermont said that the Democrats had "made a very determined effort to stir up a farm revolt," but added that the plan was not working. Attacks on Benson were bitter, and even vicious, but Eisenhower avoided the wrath of farmers. The Democrats were vexed and frustrated when Benson acted as a kind of lightning rod to draw the wrath away from the President. Eisenhower easily won reelection, although the Republicans lost a few seats in the House and Senate.[12]

As Eisenhower began his second term, agricultural surpluses continued to outrun domestic and foreign demands by huge amounts. Federal farm programs were costing the government some $2.5 billion annually. The halfhearted attempts to reduce production more in line with market demands were no match for the growing productivity on American farms. The cumulative effects of science and technology in agriculture defeated all of the policy makers' efforts to control price-depressing surpluses.

During the years after World War II, every aspect of the agricultural revolution rapidly accelerated. Major progress occurred especially in mechanizing grain and forage production. Larger tractors and bigger and better machines for preparing the soil, planting, cultivating, and harvesting came into common use. Corn planters, for example, were not often larger than four-row in the 1940s; but in the 1950s many progressive farmers moved to six- and eight-row equipment. Self-propelled harvesters for wheat, corn, and soybeans came into general use. Other machines included self-propelled windrowing equipment for harvesting, hay balers, forage harvesters, loaders, conveyors and unloaders, elevators, machines for clearing and leveling land, and many others. The distribution of electricity permitted the use of semiautomatic and automatic equipment around the farmstead. Grain and silage were moved by electric power at the touch of a switch, and automatic watering became common for livestock and poultry. Automatic grading, candling, and cartoning of eggs became standard. Milking machines and the cooling of fresh milk were common on dairy farms. Producers also used electric chick and hog brooders. A variety of sprayers and dusters were developed to spread insecticides and fungicides quickly and cheaply. Airplanes were used to distribute these chemicals over cotton and wheat fields, while rice farmers planted their crop from the air.

Crops that had resisted mechanization fell to the inventive genius of engineers and the demand of farmers for more labor-saving equipment. Perhaps the most important of these was the mechanical cotton picker. After years of experimentation, the International Harvester Company placed its first spindle-type picker on the market in 1941. Machines to strip cotton bolls from the stalk had been used in Texas and Oklahoma since the 1920s, but the invention of a machine to pick only the lint took much longer. After World War II mechanical cotton pickers and related technology for cotton production gradually reduced and ultimately eliminated hand work among larger and more efficient cotton producers. By 1963 some 72 percent of the cotton crop was machine picked. Machines were also developed or improved to harvest many other crops that had historically required a great deal of hand labor. Peanuts, potatoes, and sugar beets all succumbed to the machine. Labor-saving machines became available for about every job in the field and around the farmstead.

Mechanization of agricultural production could not have succeeded to the degree it did without the work of plant breeders. These scientists bred crop varieties that lent themselves to mechanical harvesting. For example, cotton scientists developed strains of cotton that opened more uniformly to reduce the number of times that pickers had to go through the field, and corn breeders developed plants with stronger, taller stalks, which was an important factor in efficient machine harvesting.

The breeding of more productive crop varieties and improved herds of livestock also accelerated after World War II. In the 1950s hybrid grain sorghums were introduced and within four or five years virtually all of the crop was converted to these new breeds which produced much more per acre. Better strains of wheat, rice, soybeans, cotton, and other crops were also developed. Not only did the new varieties produce more, plant scientists demonstrated to farmers how they could increase their output even further by different planting practices. Traditionally, farmers had planted about 10,000 kernels of corn per acre. By the 1950s the planting of 20,000 seeds per acre became common practice by the more efficient corn farmers. Livestock herds were improved through artificial insemination and the use of high-quality parent stock. Animal scientists worked to improve the nutrition in feeds for livestock and poultry in order to get more gain with less feed. Still others sought successfully to control plant and animal diseases.

A third aspect of the accelerating agricultural revolution centered around the wider use of chemicals for fertilizer, as well as all kinds of insecticides, fungicides, and herbicides. While farmers had been gradually increasing their applications of commercial fertilizers, mainly nitrogen, potassium, and phosphorus, before World War II, consumption rose dramatically after 1945. In the fifteen years ending in 1960, American farmers increased their use of primary plant nutrients from 2.6 to 7.3 million tons annually. Moreover, insecticides such as DDT, which was introduced in the 1940s, and herbicides and fungicides, greatly reduced crop losses from insects and weeds.[13] Airplane spraying of cotton, wheat, and other crops drastically cut labor costs for insect control. In the case of cotton, chemical defoliation preparatory to mechanical picking helped to bring about complete mechanization of that crop.

As a writer in *Fortune* magazine put it in 1953, "a chemical revolution had hit the farmer with a bang." And so it had. The big chemical firms and some of the oil companies rushed into the farm chemical business and by 1952 were selling farmers between $300 and $400 million worth of insecticides, fungicides, and herbicides each year. Many of the products that farmers were using in the 1950s were entirely unknown only a decade earlier. Change was so fast that one sales manager complained that the research department was coming up with something new and better "to obsolete a product just as we've got it half-way established on the market." Unlike the old-line arsenicals and lime-sulphur mixtures that farmers had used for years, the new chemicals were highly toxic, acted with much greater precision, and met highly specified needs. For example, the herbicides would kill weeds, but not the corn or cotton plant.[14]

A fourth aspect of changes on American farms was the growing specialization and a greater emphasis on modern management practices. Some farmers had always specialized to a degree, but by the late 1950s most commercial producers were concentrating their labor, capital, and land on one or two main crops, or one or two crops combined with a livestock enterprise. For example, the successful midwestern farmer would characteristically produce corn, soybeans, and hogs, or perhaps just corn and soybeans. Those farmers quit milking cows, tending chickens, feeding a few steers, making hay, or doing numerous other things so common on midwestern farms before World War II. Such an activity as cattle feeding, which had been done in many parts of the country earlier, became specialized on the Southern Plains and in the Central Valley of California where feedlots held as many as 100,000 head. Fewer and fewer farmers did any milking as the larger dairymen with sometimes hundreds of cows took over milk production. Overall, most of the major commercial grain, cotton, livestock, and poultry enterprises grew larger and more specialized.

Agriculture's new directions required much better management than that practiced by most pre–World War II farmers. Hard work and luck were no longer enough to avoid bankruptcy. The modern commercial farmer had a large investment in machinery, fertilizer, seed, prepared feeds, and perhaps livestock, as well as land. For the larger and more efficient producers this figure commonly reached as much as $100,000 to $200,000 by the mid-1950s. To safeguard that investment and to make a profit the farmer had to manage his financing, production, and marketing with the same skill as any other businessman who wanted to maximize his return on capital and labor. Farming for the modern commercial producer was a business that required intelligent decision making. Financial management involved such decisions as when and how much money to borrow, whether it would pay to invest in more storage capacity, or whether to buy or lease a fertilizer spreader. Production management was concerned with such matters as what crops or livestock to raise, what size machines to purchase, or how much fertilizer to apply. Marketing decisions were also important. Should a farmer contract his crop ahead of production, should he sell at harvest time, or would it be better to store the crop?

Modern commercial farmers in the United States were faced with a host of management decisions. Experience, reading, short courses, and observation were among the ways progressive farmers kept abreast with rapid change. Those who best understood how to use financing, or how to adopt the most efficient machines to their operations, and those who had keener insights into the marketing processes were the ones who had the best chances of financial success.

What were the forces behind the great changes on America's commercial farms? One of the most obvious influences, of course, was the cumulative effect of changes that had been going on for many years. But beyond this was the fact that farmers themselves wanted to increase their efficiency and production to improve their incomes. Farm families wanted to enjoy the same standards of living as people in nonfarm employment. They wanted to modernize their homes, to buy household appliances, to educate their children, and to take vacations like town and city folks. These things required more income. Since farm income was determined by units of production times price, farmers believed they could make more money by increasing their efficiency. This meant producing more crops and livestock in relation to the inputs of capital and labor. Relatively low farm commodity prices in the 1950s placed added pressure on farmers to produce more crops per acre, more milk per cow, more pigs per litter, or faster weight gains on livestock and poultry.

The USDA, the colleges of agriculture, agricultural experiment stations, and the cooperative extension services all played a significant role in helping farmers achieve their goal of increased production. While these agencies had been working with farmers for many years, their influence before World War II had been modest. Farmers had resisted "book farming," and relatively few producers looked to the agricultural colleges and experiment stations for help and advice. As farming operations became increasingly complex, more and more commercial farmers turned to scientists, engineers, economists, and other specialists at these institutions for assistance. Wheat farmers on the Southern Plains looked to Kansas State University for information on the best varieties of seed, methods for disease and rust controls, proper fertilizer applications, and other matters. Cotton growers in the South depended on the work of engineers and plant breeders at the agricultural experiment station at Stoneville, Mississippi, to develop a cotton plant and machinery which together would permit fully mechanized production. At the Tifton, Georgia, experiment station, extensive research on the peanut in the 1950s resulted in farmers doubling their production per acre. Farmers no longer ignored the information streaming from the agricultural colleges. As one large Alabama cotton grower said in 1962, he had been using preemergent weed control ever "since it was first recommended by Auburn."[15] Nutrition scientists at the agricultural colleges did research on improved feed for poultry and livestock, while other researchers showed farmers the best methods for disease control.

Farm machine companies, chemical firms, and other industrial corporations which sold supplies to farmers also encouraged new trends in agriculture. Indeed, farmers came to depend on the local farm machine dealer, the distributors of fertilizer and other chemicals, and the feed and seed dealers for production advice. Representatives of these groups urged farmers to buy the latest model tractor, feed that stimulated rapid growth of livestock, and fertilizer to increase crop output. Businesses that supplied the increasing amounts of nonfarm inputs for farm production had a high stake in the dominant agricultural trends. Through demonstrations and advertising, they unceasingly urged farmers to use their products.

Under the impact of farm technology, plant and animal sciences, and modern farm management, commercial agriculture made remarkable productivity gains after 1945. While total acreage harvested actually declined from 354 million in 1945 to 324 million in 1960, the output of crops rose about one-fourth and livestock production rose nearly 20 percent in that fifteen-year period. In some cases and in particular areas production rose much more dramatically. For example, corn averaged between 35 and 40 bushels to the acre in the 1940s; by the early 1960s the average was more than 80 bushels in Iowa and Illinois. In 1945 the nation's wheat crop on 65 million acres reached 1.1 billion bushels, but by the late 1950s production was as large or larger on 10 to 15 million fewer acres. In the same period soybean output rose from only 193 million to around 550 million bushels.

While total production increases by American farmers were unparalleled, the most distinctive development was the rise in output per man-hour. In 1945, 18.8 million man-hours were devoted to farm labor, while in 1960 this had dropped to only 9.8 million. Farmers were producing much more with about half as much labor. This had been made possible by substituting capital for labor. By the 1950s and 1960s agriculture had become a highly capital-intensive industry. Each American farm worker by 1960 produced enough agricultural products for 26 other people.

The increased efficiency of labor in agriculture permitted individual operators to handle much more cropland and livestock. Indeed, farmers needed more land in most cases to use agricultural technology effectively. Consequently, the size of farms rose rapidly after 1945, and the number declined sharply. On the eve of World War II there were still more than 6 million farms in the United States. By 1960 the number had fallen to 3.9 million. Most of the commercial production came from only about 750,000 of these farms. Between 1940 and 1960 the average farm had grown from 174 to 303 acres; the people living on farms had declined from about 30 million to 15.8 million. In those two decades, farmers dropped from 23.1 to 8.9 percent of the total population.

But with growing efficiency and increasing productivity down on the farm, fewer operators continued to produce huge surpluses in the 1950s. A generally prosperous domestic economy, free distribution of food to school lunch and relief programs, shipments of large quantities of grain to starving people overseas, and regular foreign markets were not enough to use the production of American farmers—not even when combined with some modest acreage restrictions. This did not mean that everyone had enough to eat, but it did indicate that, at even relatively low prices, there was much less effective demand for farm commodities than there was supply.[16]

The resulting unfavorable supply/demand situation created a cost-price squeeze that forced tens of thousands of farmers to tighten their belts, seek part-time nonfarm work, or leave farming altogether. As Winifred Horner, a young farm wife from near Columbia, Missouri, wrote in early 1956, "in the midst of an abounding prosperity, we are facing our own private depression." In some respects, the David Horners were a typical young farm family in the Midwest who had been farming for eight or ten years. Horner had some advantages in that he had a degree in agricultural engineering and began farming in 1947, a favorable period. After two years as a tenant and with some help from the Veterans' Administration, the Horners acquired a 245-acre farm and rented additional land which gave them a 600-acre operation. Besides crop farming, for which they gradually built up a $6,000 line of machinery, they developed a fine cattle herd.

In their early years things seemed to go well. They did not make much money, but they built up their capital in land, machinery, and livestock. In the fall of 1951, Mrs. Horner wrote, "we felt confident we were operating on a safe margin and could withstand a normal weather hazard or a reasonable drop in the market," conditions fairly common for farmers. They were not prepared, however, for what followed. In 1952 they lost $4,200. Their cows that had been worth $300 a head in 1951 dropped to $160 in September 1952, and by 1956 they would bring only $110 per head. Other farm prices also declined sharply, and then in 1954 drought struck. Horner found that his farm income was not enough to cover his operating expenses, payments on his debt, and leave enough for living expenses. While city people often viewed farmers as fairly self-sufficient, by the 1950s they were not only dependent for off-farm inputs for their production, but bought items for daily living just as their friends in town did. The Horners had to pay telephone and electric bills, they bought groceries at the supermarket, faced fuel bills for the car and tractor, and had educational expenses for their children. These were not occasional expenses, but had to be met regularly from a shrinking farm income. Hard work and better than average management were not enough to solve the Horners' cash flow problem.

In order to make ends meet Horner took an off-farm job. Low farm prices and high prices for the things they had to buy, Mrs. Horner wrote, were hazards peculiar to the agricultural economy in the mid-1950s. A farmer could not pay for a $3,000 tractor with 95-cent corn. For those out on the farm, the cost-price squeeze was not merely something academic that agricultural economists calculated and politicians talked about correcting in a search for votes. It was a crushing burden which not only eliminated the weak, but forced many farmers who appeared fairly successful into a tenuous position. "We have seen our contemporaries leave, one by one," Mrs. Horner wrote about her community. "Some go fast, in a cloud of disgust, selling everything behind them. Most ease out slowly. First they get a good job off the farm. Then the farm work comes to a standstill because they don't have time for both. Then it's lonesome for the family, and pretty soon they are looking at a ranch house in the suburbs."[17] This, indeed, was the pattern ultimately followed by the Horner family.

This experience, shared by thousands of family farmers like the Horners, was not confined to the period of the mid-1950s, nor to any particular section of the country. Whenever agricultural prices dropped in relation to the cost of operation, many farmers were placed under severe financial pressure. These conditions, of course, did not affect all farmers in the same way or cause equal distress. Young farmers who usually had long-term debt for land, and shorter-term obligations for machinery and operating expenses, were hardest hit. If a farmer went into debt for land and machinery in 1947 or 1951 when agricultural prices were relatively high, actually above 100 percent of parity in some cases, and then tried to pay off these debts with income from products whose prices might drop 10, 25, or even as much as 50 percent, the debt became a very crushing burden. It was even worse when the prices of tractors, fuel, fertilizer, seed, and other operating expenses remained high. Throw in some bad weather which reduced production, the loss of livestock from disease or some other calamity, and farmers faced real tragedy. Under these circumstances, farmers tried to reduce their expenditures by getting along with less, extending their payments (which only added to interest costs in the long run), taking part-time work off the farm, or in extreme cases just leaving the farm altogether.

As farmers became more commercialized, they were absorbed completely in the cash and credit economy. They had to produce cash income to meet the expenses of nonfarm inputs and living costs. As one young Iowa farmer said in 1956, "We buy our living on farms today. Twenty-five years ago our dads used to produce it right on the farm." As late as the 1920s, and even into the 1930s, most farmers raised their own power—horses and mules and oats and hay. Power was not much of a cash expense. The same was true of fertilizer, which was mainly barnyard manure, or seed or feed, which came from the farmer's own fields. But by the 1950s these were all cash items. When the tanker operator filled the gasoline storage tanks to fuel the tractor, when the dealer delivered commercial fertilizer, and when the seed salesman sold seed, they all expected payment. Sometimes local businessmen extended short-term credit, but the farmer had to generate income to pay those bills. The same was true of machinery repairs, garage work, and a host of other expenses. Also farmers were trying to raise their standard of living by improving their houses, installing modern plumbing, buying automobiles, radios, television sets, refrigerators, freezers, and other household items. It all took money, and a lot of it.

Older farmers were in a better position to meet a cost-price crisis. Many of them, after somehow weathering the Great Depression, purchased land in the late 1930s when it was literally "dirt cheap." In east central South Dakota, farmers bought land for as little as $8 to $10 an acre in 1938 and 1939, and in Iowa good land could be purchased for as little as $50 to $100 an acre. The farmer who owned land on the eve of World War II was, as it turned out, downright lucky. Prices began to rise and stayed relatively high for more than a decade. The period from 1940 to 1952 was the longest period of sustained prosperity in American agricultural history. Many farmers bought land in 1939 and 1940 and had it paid for by the end of the war. Operating with cheap land and fairly low-priced machinery, these farmers did very well during the war and postwar years. They got out of debt and put money in savings, or more land.

When the price breaks of 1953 and 1954 arrived, these farmers were hurt, but not seriously. They had no debt payments, they had enough land and production to qualify for substantial cash benefits under the federal programs, and they had already purchased many items for modern living. They just had to earn enough income to pay their operating and living expenses. It meant some belt tightening, but no real change in living standards or threat to their basic economic position. Some farmers may have actually benefitted from the hardships imposed on other farmers because they were able to buy additional land at profitable prices from those neighbors being driven out of farming by the cost-price squeeze. These financially solid farmers thus acquired more land and reduced unit costs by increasing their efficiency and productivity. This placed even greater competitive pressure on the smaller, less efficient producer and contributed to the rapidly declining number of farmers. With more land, the larger farmers were the chief beneficiaries of the federal agricultural programs. Thus, even in the mid-1950s, when the cost-price squeeze hit the Horners and tens of thousands like them, other farmers were making a comfortable living, even expanding their operations and building up a substantial estate. While younger farmers with less land and limited capital were hit the hardest, some young operators did well. Some had family backing that gave them capital and a degree of security against price drops, and others were raising the right crops, had good weather, were close to market, or enjoyed other advantages. Luck was not unimportant.

Overall, economic conditions placed such heavy financial pressure on marginal and near-marginal farmers in the 1950s that many of them went out of business. The cost-price squeeze was a major factor in driving tens of thousands of farmers out of agriculture and in contributing to the growing concentration of land in the hands of fewer and fewer operators. In 1959 there were 1,677,934 fewer farmers in the United States than at the beginning of the decade. This loss of about 167,000 farms annually disturbed many people, both in and out of government. While many observers were not optimistic about the future of small farmers, most Americans seemed emotionally tied to the desirability of maintaining the family farm. Whenever a farm was absorbed by another farmer and the family moved away to start life over somewhere else, it appeared to many people that another little bit of the true American heritage had disappeared. Furthermore, there was growing criticism of large farmers, corporation farms, and agribusiness. These factories in the fields, critics said, were driving the ordinary family farmer off the land. What, if anything, could be done to help the hundreds of thousands of low-income farmers? That was a common question in the 1950s, and an unsolved problem.

Poor Farmers, Agribusiness, and the Family Farm

VII

As farmers, farm organization leaders, and Congress argued over production, markets, prices, and farm policies in the decade after World War II, little was said or done on behalf of farmers at the bottom of the economic scale. The concern expressed for millions of poverty-ridden farm residents in the days of the Resettlement Administration and the Farm Security Administration largely faded under the impact of war and the indifference of agriculture's power structure.

General wartime prosperity turned peoples' attention away from the problems of the rural poor, while industrial jobs and military enlistments drew many low-income farmers completely out of agriculture. Between 1940 and 1950, for instance, 122,700 generally poor black farmers in the South left agriculture for other employment. The lack of nonfarm jobs which had held people on the land during the Great Depression suddenly disappeared. Now alternative opportunities that historically had been open for the surplus farm population appeared once again. Hundreds of thousands of farm people finally had a chance to abandon their stark existence.

The wartime and postwar prosperity, however, largely bypassed those low-income farmers who continued in agriculture. Indeed, very little changed for them in the 1940s. High prices meant nothing to the small operator who had little or nothing to sell; increasing land values were of no benefit to the landless sharecropper or tenant. The 1950 census revealed that more than 700,000 farmers, or 13 percent of the total, sold only between $250 and $1,199 worth of products in 1949. Even though many of these farmers had a few days of off-farm work and were classified as commercial operators, most of them were extremely poor. Another 901,318 sold farm commodities valued at only $1,200 to $2,499. Of the 5.4 million farms in 1949, only 2.1 million had sales exceeding $2,500. But net income was what counted. A study of conditions in 1950 showed that about 1.5 million families on small farms, or 28 percent of all farmers, had a net cash income of less than $1,000 from all sources, including off-farm work.[1]

The fact was that some sections of the American landscape were inhabited by tens of thousands of poverty-stricken farm people at a time when most of the country was enjoying a burgeoning prosperity. The Jeffersonian ideal of a happy, independent, comfortable existence on the land was not even a dream for thousands of rural poor. As one observer explained, it sounded romantic to talk about farm life, "with its concomitant evocation of lush pastures dappled with white-faced cows, and kitchen gardens sparkling with dew on June mornings," but this picture, he wrote, was the image not the reality. The true condition for many farmers was that they lived in leaky, drafty houses with no modern facilities, existed on inadequate diets, and were denied education and health care. By 1950 there were around a thousand counties in the United States where, as one federal investigation found, "more than half of the farmers are mainly dependent on the income from small, poorly paying farms."[2]

Most of the poorest farmers did not live in poverty because of low prices or the cost-price squeeze, which concerned the more successful commercial producers. These poor farmers faced much more basic problems. They lacked enough good land, they had no access to capital or credit with which to buy more land or modern equipment, and many of them lacked education and management skills that were becoming increasingly necessary in business-oriented agriculture.

What was the answer to the problem of rural poverty? Or were there any good answers? Clearly, the programs designed for the larger commercial producers held no promise for the lowest economic classes in agriculture. The Farmers Home Administration, successor to the Farm Security Administration, made land-purchase, production, subsistence, and disaster loans to low-income farmers, and provided some supervision and management. However, in the late 1940s and 1950s the Farmers Home Administration seldom made as many as 5,000 farm-ownership loans a year in the entire country. At that rate it would take two centuries to assist the poorest farmers to become landowners! From 1937 to 1951 the federal government loaned $427 million to more than 60,000 farmers to buy farms or to increase their small holdings. That represented a mere 3 percent of the farm tenants in the nation. Most of the poorest farmers did not even qualify for this kind of credit. It was clear that no existing government policies were reaching the most needy farmers or helping them to become viable commercial operators.[3]

In the 1940s the question of poverty among farmers was mentioned from time to time in the popular press and occasionally came in for serious study, but in general the subject was avoided. Most people seemed to assume that somehow the wartime prosperity was trickling down to the rural disadvantaged. Whatever was said about poor farmers was drowned out by the fight of commercial farmers for parity prices. Early in 1950 Theodore W. Schultz, one of the nation's leading agricultural economists, surveyed the problem of what he called "poverty within agriculture." Schultz emphasized that the "chronic problem of poverty . . . in many parts of agriculture" was viewed by most people as something "private and personal" and having no "social roots or major social implications." Not all farmers were poor, Schultz stressed, but the agricultural programs did not deal with the poverty that did exist. Agricultural poverty, he said, was viewed as something natural and to be expected. People easily accepted the idea, for example, that blacks and Chicanos were supposed to be poor. The situation of poor farmers had not been studied sufficiently, according to Schultz, because they were "politically impotent."[4]

One question raised in connection with the poorest farmers was whether they should somehow be helped to continue farming or urged to seek employment elsewhere. Most agricultural economists believed that the nation had too many farmers, that there was a surplus of people in agriculture. With an excess of labor in relation to the resources of land and capital, these authorities argued, real farm relief for most small farmers was somehow to get them completely out of agriculture, or at least provide opportunities for part-time nonfarm employment. This position reflected the belief that there was no economic future for the great majority of people who tried to make a living on a small farm. The quicker these farmers could find full-time or part-time off-farm work, the better off they would be. To facilitate out-migration from agriculture, it was suggested that the federal government establish a national employment service, set up programs to give special training for nonfarm jobs, and make loans to families to help them move to localities where work existed. In other words, government policy should be directed toward improving the economic mobility of poor farmers so they could leave the farm.

Most Americans, however, did not favor government policies that would accelerate the decline of the farm population. Such a policy ran counter to the deep feelings held by many people that there was something good about life on the land, both for individuals and for the nation. While about half the farmers lived a life that was a far cry from the Jeffersonian dream, the nonfarm population had little understanding of the economic and social problems faced by the poorest classes of rural residents. Town and urban people clung to an image that varied greatly from actuality. Farmers themselves opposed any government effort, as one writer put it, to "get rid of marginal farms." When queried in 1953 about this approach to curing farm ills, 62 percent of the farmers rejected the idea of any government policy that would encourage or force people off the farm. Even if farmers were inefficient or maintained a bare subsistence, nearly two-thirds of the respondents believed that forcing farmers to leave the farm was un-American and even non-Christian.[5] Consequently, little support could be generated for policies that would consciously move people out of agriculture. But unless low-income farmers could be helped to become successful producers, or unless they left agriculture for other employment, hundreds of thousands of rural families would be committed to permanent poverty. Planning in the agricultural sector was not equal to solving this cruel dilemma in the post-World War II years. Government chipped around the edges of the problem, but never faced it directly until the 1960s.

Whatever the government did or did not do, it seemed certain by the late 1940s and 1950s that the decline in the number of farms and farmers was irreversible. No amount of oratory detailing the economic, political, social, or spiritual values of farming and farm life could affect that trend. Some poor farmers, but usually not the poorest, could be and were helped by the Farmers Home Administration and other government agencies to become successful operators. More low-income farmers might have been assisted if Congress had funded some of its programs more generously. Long-term, low-interest credit for land and machinery, the purchase and improvement of land for sale to needy farmers, adjusting parity prices in favor of smaller producers, government support for cooperatives, and money for rural housing were among the suggestions made to alleviate rural poverty. Most lower-class farmers, however, could not be transformed into productive agriculturalists even with the expansion of such programs.

The idea of establishing farmers on their own small acreages was no longer an answer to the problems facing so many rural residents in the 1940s and 1950s. Unless the revolutionary changes occurring in American agriculture could be stopped, to place a family on 50 or 75 acres of land was, in most cases, to commit it to a low income forever. Under modern agricultural conditions and the rising cost of living, there was, with few exceptions, no way that a family could prosper on production from such small operations. The need was for a better balance between human resources and land resources. The productivity of many farmers was simply too meager ever to provide a decent living. However unpleasant the truth might be, thousands of farmers were not needed in agriculture. To try to keep them on the farm might have satisfied a deep emotional desire, but could not be justified in economic terms. One reason Secretary Benson was so unpopular among some groups in the 1950s was that he openly advocated getting the poorest farmers into nonagricultural employment. The solution to conditions among poor farmers could not be found in the agricultural sector but needed to be sought off the farm. As Herschel D. Newsom, Master of the National Grange, told the annual assembly in 1954, it would never be possible to completely meet the problem of low-income farm families "within the business of agriculture." The question that faced both individuals and the nation as a whole was how best to transfer these unfortunate farmers into other lines of work where they could be fully employed and could increase their productivity.

With developing mechanization and other modern agricultural practices, there was no economic reason to have so many farms and farmers. The main problems facing commercial producers in the 1950s were surpluses and low prices for farm commodities in relation to the prices they had to pay for industrial goods and services. Government programs of acreage restriction to reduce output in order to maintain fair farm prices contradicted any policy of trying to help the marginal farmer, who usually needed more land to make a decent living. A Georgian wrote Senator Richard Russell in 1954 that several farmers in his community had left their farms to seek other work because they did not have enough acreage on which to earn a livelihood. Reducing the total amount of cultivated acreage at the same time that machinery was becoming larger and more efficient could only mean, as Walter Lippmann observed in 1956, that "several million farmers will have to get off the farms." To deal humanely with the needed change, Lippmann argued, something needed to be done "to slow up, to soften, to alleviate this painful human readjustment." A writer in *Fortune* said that "doing something about redundant, underemployed 'farms' and 'farmers' has become a national issue," and he recommended that the federal government should help the least efficient, low-income farmers to leave agriculture. However, the editor of the *Daily Oklahoman* in Oklahoma City wrote that no politician or political party would dare to write "a farm evacuation plank into its platform."[6] That was certainly recognizing political reality.

While the problems of commercial farmers were the center of attention in the 1950s because of surpluses and low prices, in January 1954 President Eisenhower called for a study of "the problems peculiar to small farmers," most of whom were not in the commercial mainstream. Some fifteen months later Secretary Benson submitted a study done by the United States Department of Agriculture entitled "Development of Agriculture's Human Resources." In submitting the study Benson told the President that "clearly, a broad, aggressive, well-coordinated assault is urgently needed" on the problems of rural poverty. While the investigators found that the number of low-income farmers had been declining in absolute terms, the trend, they said, needed to be accelerated. They emphasized what many had been saying, both in and outside of farming, that the "solution to the problems of farmers with low earnings lies outside commercial agriculture." This meant getting farmers completely off the land, as well as increasing the number of part-time jobs.

The USDA study did not make any startling recommendations. Whatever was done, the report stated, must be "within the American philosophy that each individual make his own decisions and set his own goals." In practical terms this meant that the government would offer various types of assistance to some lower-income farmers who were trying to develop a viable farming operation, but that the long-range solution to poverty in agriculture would be found in full- and part-time nonfarm employment. At the same time, the study recommended such things as more credit, technical assistance, vocational training, the adaptation of agricultural extension work to low-income farmers, Farmers Home Administration loans to part-time farmers, and establishment of industry in rural areas to provide jobs for underemployed farmers.[7]

Neither at the time nor over the next several years, however, did anything substantial result from these recommendations. In 1955 the Eisenhower administration asked for special funds to begin fifty rural development pilot projects, but Congress refused to make the necessary appropriations. Even though no meaningful funding was forthcoming, Secretary Benson did inaugurate a small program to help start rural development committees. These committees, with the assistance of established federal agencies such as the Employment Service of the Department of Labor and the Farmers Home Administration, sought to improve rural economies in some depressed areas. By 1960 there were projects in thirty-eight states. These included starting a few small industries to provide off-farm work to underemployed farmers, forming cooperatives, and improving production and marketing of specialty crops. But in the late 1950s Congress was in no mood to make a formidable attack on rural poverty. The political pressures generated at the grass roots continued to emphasize parity prices, not relief or reconstruction for poverty-ridden rural residents. Schultz was right: the issue of rural poverty remained unsolved because those involved lacked political power. Consequently, for the most part, poor farmers continued in poverty.

At the bottom of the agricultural ladder were some 300,000 to 500,000 migrant farm workers. Originating mostly in the South and Southwest, these laborers moved with the production cycles of fruit, vegetables, sugar beets, and other crops that required intensive seasonal hand labor. Some migrants were displaced farmers. Migrant workers suffered from extremely low incomes and eked out an existence on only a few hundred dollars a year. They ate poor food, lived in extremely bad housing, and suffered from an absence of educational opportunities and decent health care. In the 1930s, John Steinbeck and other writers had dramatized the condition among these unfortunate rural residents. However, during the 1940s and 1950s nothing much was done to alleviate their situation. Presidential commissions on migratory labor were established, which made studies and reports, but remarkably little changed in the working or living conditions of farm migrants.

While the poorest farmers received only token recognition, by the 1950s the increasing number of large farms aroused deep concern among many Americans. Every new statistical study showed that large corporate and integrated producers were growing bigger and playing an increasing role in the nation's agriculture. Some observers believed that these huge operators were a threat to agriculture's middle-class family farmers and therefore a danger to the country. For many years, of course, large farms had characterized part of American agriculture. In the late nineteenth century the western ranches and the bonanza wheat farms of North Dakota and California had typified corporate farming. The corporation farm was not only a special legal entity, it was a business in which the management and ownership of land were separated from hired labor. In this respect it was different from the family farm where members of the family furnished most or all of the labor and provided the management.

In the twentieth century, the corporate farm took over the production of much of the fruits, vegetables, nuts, and some other specialized commodities. These large operators used industrial methods of production and often gained control of processing and marketing as well. The corporate structure, however, had not proven especially advantageous in most aspects of grain, cotton, or livestock farming. The statistics on corporate farming were inadequate, but certainly by the 1950s there were only a few thousand corporate farms in the United States. They probably produced less than 5 percent of the farm output. Most of these corporations were highly specialized and were located in California, Arizona, south Texas, and Florida.

While corporation farms were not numerous, their existence and growth raised the specter of monopoly and unfair competition in agriculture. Some states had laws prohibiting corporate farming and others placed statutory limits on how much farmland could be owned by a corporation. But no amount of criticism could stop the growing number of large farms, both incorporated and unincorporated. When judged on the basis of farm size, amount of investment, or products sold, fewer and fewer of the largest farmers were becoming more and more important in the nation's agriculture.

In 1950 the census reported nearly 5.4 million farms. Of these, 3.7 million were considered commercial, and the other 1.7 million were labeled "other farmers." In this last category there were 630,230 part-time farms where the operator worked in nonfarm employment 100 days or more a year, and 1,029,392 rural residences. The commercial farms were organized into economic classes on the basis of the value of products sold in 1949. Class I farms had sales of $25,000 or more, while Class VI farms sold farm products worth between $250 and $1,199 and the operator worked less then 100 days off the farm. The other classes fell between these extremes. It was clear that a vast gulf existed between the condition of Class I and that of Class VI commercial farmers.

Between 1949 and 1954 the number of Class I farms rose from 103,231 to 134,003. While the Census Bureau changed the definition of a Class I farm in 1959 to one selling $40,000 worth of products or more, large farms continued to grow. More than 102,000 farmers sold products in excess of $40,000 by 1959. Although only 4.2 percent of all commercial farmers fell in this category, they harvested 15 percent of the cropland and produced 32.8 percent of the value of all farm products sold. When Class II farmers are added, those whose sales ranged from $20,000 to $39,999, the two top groups in 1959 made up 12.9 percent of all commercial farmers, farmed 32.3 percent of the cropland, and produced 51.9 percent of the value of all farm products marketed.[8] California, Texas, Iowa, and Nebraska ranked in that order in the number of Class I farms by the end of the 1950s. California and Texas had a substantial number of corporate farms, but the great majority of these big operators represented large, rather elite family-run enterprises.

To some observers the greatest danger to traditional American agriculture was not the huge family farms, even if they were incorporated, but the growth of vertical integration and contract farming. By the 1950s an increasing amount of agricultural commodities was produced by firms that controlled food output from producer to consumer. This "vertical integration" was achieved by gaining control of the production, processing, and marketing of certain farm products. In 1955 John H. Davis, a former Assistant Secretary of Agriculture, coined the word "agribusiness" to describe this development. "Agribusiness," Davis said, "refers to the sum total of all operations involved in the production and distribution of food and fiber."[9]

Contract farming, of course, was not new. It had been employed in fruit and vegetable farming and in some other aspects of agriculture for many years. Under this system, for example, the Campbell Soup Company would make contracts with vegetable growers in New Jersey for the delivery of so many tomatoes at a fixed price. The company would then process the tomatoes into soup and distribute it to stores and supermarkets. In California, Sunkist Growers, Inc., a federated cooperative, contracted with citrus growers for the delivery of oranges and grapefruit. Sunkist, which had its own trademark, then integrated the picking, packing, and marketing of the products. In the sugar-beet and sugarcane industries, practically all growers had contracts with sugar companies for the delivery of their output.[10]

In the 1950s integration moved into new segments of agriculture. One of the best examples was poultry. Feed companies, such as Ralston Purina, made contracts with farmers to furnish chicks, feed, medicine, fuel, and any other needed supplies. The farmer only provided the broiler or laying houses, equipment, and labor. If it were a broiler operation, the contractor received the broilers and paid the farmer for his labor—usually so much per chicken. While there was practically no integration in the poultry industry in 1949, between 1950 and 1960 the great majority of the broiler business came under control of integrated operators. By 1963 contract growers controlled some 75 percent of the broiler industry and 20 percent was in the hands of company-owned broiler farms. Poultry raising was no longer a part of general farming operations by 1960.[11]

Several factors were at work to bring about the growth of agribusiness. In the case of chickens, independent producers of broilers and fryers had found the cost-price squeeze becoming tighter and tighter. Many of them went broke in the 1950s as prices fluctuated and profits vanished. But more important, as the country became more urbanized and the mass market for food increased, a need existed for the regular delivery of huge quantities of agricultural products. In short, urban consumers required a reliable supply of high-quality farm products. Moreover, as more women worked outside the home, they demanded prepared and semiprepared foods. There was a direct connection between women's liberation and the food services provided by agribusiness. The agricultural marketing system that had served the nation earlier was not adequate to meet the new demands. For example, hundreds and thousands of dairy or poultry producers could not individually provide uniform quality or reliable delivery of their products to consumers without some central or integrating agency. The agribusiness companies, or the farmer cooperatives, that integrated the functions of production, processing, and distribution to retailers fulfilled a useful economic function.

Agribusiness companies, of course, were in business to make a profit, and they usually made more money by integrating all stages from production through consumption. In the poultry industry, for instance, the feed companies sold more feed as they contracted with farmers to grow more broilers for a year-round market. Greater volume reduced unit costs and increased profits. As one authority wrote, "the incentive to integrate is based on volume, continuous flow, and low unit costs."[12] While agribusiness companies reaped the main benefits from integration, some advantages accrued to farmers. They had an assured market, their risk was reduced to some degree, and the agribusiness firms often advanced capital for expansion and supplied technical guidance and advice. The picture sometimes presented of a grasping agribusiness firm greedily driving a happy, prosperous, independent farmer out of business in order to increase its own profits was not unknown, but it was not necessarily typical. In the case of poultry farmers, many of them had already gone broke before the agribusiness firms moved in.

Critics of integration, however, charged that agribusiness gained too much power over farmers and farming. Under the system, farmers lost their freedom to make management decisions and for all practical purposes became hired workers rather than independent proprietors. It was said that producers were at the mercy of the multimillion-dollar corporations which established the production and prices of many commodities. Once a farmer became part of an agribusiness organization he lost his independence and became only a link in the powerful industrial food and fiber chain.

The expansion of agribusiness and corporate farming in the 1950s aroused deep concern among many Americans. If the trend continued, some observers asked, would the end result be a decline and perhaps even the ultimate destruction of the family farm? As the editor of the *Daily Republic* in Mitchell, South Dakota, asked in December 1957, would a handful of large corporations come to control agriculture while the rural towns and cities wasted away because of lack of farm customers? This was a nagging question for which no one seemed to have a good answer. It was clear in the 1950s that many middle-class farmers were suffering from a severe cost-price squeeze. The number of middle-income commercial farmers whose total sales averaged between $2,500 and $9,999 dropped by more than 50,000 in the five-year period from 1949 to 1954. At the same time the number of large farm enterprises rose by more than 30,000. In 1956 AFBF President Charles B. Shuman told the Farm Bureau's annual convention that agriculture was still "predominantly a family business," but he added that farmers who "insist on using outmoded and inefficient methods will be eliminated from farming."

The issue was how many farmers in the years ahead would be unable to meet the high standards of efficiency required in modern agriculture and stay in business. Even with greater efficiency, there was a question of whether the family farmer could compete with agribusiness corporations. It was one thing for farmers to fail because of poor management or bad luck; it was something else for them to be forced out of business by a corporate farm backed by millions in capital made in industry, commerce, or finance.

The prospect of a continuing decline in the number of family farms struck at some very deeply held American emotions and beliefs. Indeed, the family farm was at the center of the agrarian tradition, which claimed that agriculture provided a superior way of life, as well as a good way to make a living. There was scarcely any institution in American society that had stronger emotional support than did the family farm. It ranked with the school and the church. Family and farm were both hallowed institutions, and combining them into one concept produced a powerful allegiance. Although there were many definitions and images of the family farm, most people viewed it as one where the owner or independent tenant provided all or most of the labor and made the production and marketing decisions for the farming enterprise. Throughout American history these family-type farms varied in size, kinds of crops and livestock raised, and covered a wide range of conditions from high prosperity to stark poverty. But however much they differed, individualistic, independent farmers were to many Americans a symbol of democratic government, social stability, and agricultural abundance. The family farm was a treasured institution, and any threat to it seemed to many people like an attack on the nation itself.

Concern for protecting and preserving the family farm became more intense as the number of farms and farmers declined. Strengthening the family farm received strong support from people in all aspects of American society. Agricultural economist Theodore Schultz was quoted in 1947 as saying "the family unit is traditional," and it could be justified "out of the predominant values of our time."[13] Both major political parties paid regular homage to the importance of agriculture, but the goal of protecting and fostering the family farm was a stated platform goal for the first time in 1936. The Republicans declared: "Our paramount object is to protect and foster the family type farm, traditional in American life." In 1940 both the Republicans and the Democrats expressed commitments to help the family farmer. The Democrats pledged to "safeguard the family-sized farm in all our programs," while the Republicans declared that benefits derived from government programs should be mainly restricted to "operators of family-type farms." In succeeding presidential elections family farmers were promised help and support by the leaders of both parties.

The major farm organizations paid constant tribute to the family farm and its social, political, and economic virtues. In stating its principles and policies in 1956, the Grange said that it was primarily concerned "with the family-type farm" and insisted that "room must be preserved in American agriculture for families to engage in gainful work as their own masters, on economic-sized units of land, and . . . be able to make a decent living justly comparable with that of any other part of our society." Although its leading spokesmen viewed the family farm somewhat differently, the Farm Bureau was equally committed to maintaining family farms. Of the three major farm organizations, the Farmers Union and its energetic president, James Patton, spoke most eloquently on behalf of preserving and protecting the family farm. The Farmers Union stressed that the family farm was both a business and a home, and that its loss would weaken the national character.

Presidents, members of Congress, writers, educators, and even businessmen added their praise for the family farm. Writing in the *Country Gentleman* in 1953, John Bird explained that it could be taken "for granted that this country *wants* to continue its unique family system of farming." Judged on the basis of political oratory, this seemed to be the case. In 1956 President Eisenhower expressed a widely held view when he told Congress that "more than prices and income are involved" when determining agricultural policy. "American agriculture," he said, "is more than an industry; it is a way of life. Throughout our history the family farm had given strength and vitality to our entire social order. We must keep it healthy and vigorous." In the presidential campaign of 1960, urbanite John F. Kennedy declared: "the family farm should remain the backbone of American agriculture. We must take positive action to promote and strengthen this form of farm enterprise."

In June 1956, John D. Black, a farm economist and a careful student of agriculture since the 1920s, wrote about "The Future of the Family Farm," in the *Yale Review*. After reviewing recent increases in the number of larger farms and the decline in total farm families, Black concluded that economically the trends were "in general good." But, he added, the main objections to current developments in farming were not economic in nature. He explained that criticisms "come from those who believe that living on a farm, and being reared on a farm, has [*sic*] social, cultural, and even political values so large that they outweigh what can be bought with appreciably higher incomes." As an economist, Black did not think it was wise to try to maintain people on the land who had no hope of earning a decent living. As he put it, "we can well afford to have a million or two fewer familes on farms if those we have left can live adequate lives."

The question of maintaining the family-type farm became a matter of intense attention in Congress during the mid-1950s. In late 1955 Harold D. Cooley, chairman of the House Agriculture Committee, set up a subcommittee on family farms. The purpose of this new subcommittee, Cooley explained, was "to make a special study of the ways and means to protect, foster, and promote the family farm as the continuing dominant unit in American agriculture." Congressman Clark W. Thompson of Texas, chairman of the family farms subcommittee, held hearings in late 1955 and early 1956 at locations scattered from Texas to Virginia. Thompson said his subcommittee would listen to grass-roots farmers and see what could be done to promote "the productivity and prosperity of the family operated farm, in the best interest of the institutions and economic well-being of all of us, rural and city people alike." He added that the subcommittee was "dealing here with one of the most important problems in our democracy."

Farmers were not reticent in describing their problems to visiting congressmen. Two overriding difficulties emerged from the testimony. In the first place, witnesses confirmed what all of the studies had shown for several years, namely that thousands of farmers did not have enough land or sufficient production to earn an acceptable standard of living. Acreage reduction programs actually added to the problem. Congressman Thomas G. Abernethy, a subcommittee member from Mississippi, said that at least partly because of acreage restriction programs, cotton farms in his district had an average allotment of only 13 acres. Secondly, farmers suffered from an unequal bargaining position in the economy. "Why does the farmer have to pay a fixed retail price for the things that he has to buy," asked one farm witness, "and take whatever he can get for the things he produces?" The commissioner of agriculture in North Carolina said that farmers "see other people enjoying the greatest purchasing power they have ever experienced, while farmers must continue to fight the cost-price squeeze." While the subcommittee did not offer any new approaches to preserving the family-type farm, it did conclude that "America cannot now afford to allow the substitution of a hired hand industrial type of agriculture for the independent farm family on the land."[14]

The family farm concept continued strong in Congress. In 1960 general farm legislation was called the Family Farm Income Act. According to the bill's preface, it was the "declared" policy of Congress to promote and perpetuate "the family system of agriculture against all forms of collectivization . . . in full recognition that the system of independent family farms was the beginning and foundation of free enterprise in America, . . . that it holds for the future the greatest promise of security and abundance of food and fiber and that it is an ever-present source of strength for democratic processes and the American ideal." Congressional declarations, however, could not stop the economic forces that were changing the face of American agriculture and reducing the number of family farmers. By 1960 the number of commercial farms stood at 2.4 million, a loss of some 911,000 since 1954. Even though most of these were still family-operated enterprises, the trend toward fewer and larger farms was accelerating.

The effects of a declining farm population went far beyond changes in the agricultural sector itself. Rural society and institutions were inevitably affected adversely. Indeed, by the 1940s and 1950s, churches, schools, and businesses in many rural areas had lost their base of farm support. Small-town grocery stores, grain elevators, cotton gins, farm implement businesses, and petroleum suppliers were struggling to survive in the 1950s—if they had not already closed their doors. Automobiles and trucks made it easy for the farmers who did remain to drive to larger trade centers where markets and shopping were better than in the small town. Hundreds of small towns all over the country reflected a declining rural America. Their unpainted buildings, vacant stores, and weed-filled lots presented a depressing picture and one that symbolized the final triumph of industrialization in all major aspects of American society.

Defenders of rural America, however, did not stand by and watch the decline of agricultural institutions without protest. As mentioned earlier, critics of the Eisenhower-Benson soil-bank plan attacked the legislation because they believed it would contribute to depopulating rural America. Congressman George H. Christopher of Missouri complained that the soil bank could only help what he called the "corporation type farm." The ordinary family farmer did not have enough land to retire to receive any substantial benefits. In 1956 Christopher said that during the previous three years some counties in his district had lost as many as 200, 300, and even as high as 616 farmers. Christopher was concerned that fewer farmers would mean ruin for the small businesses and professional services in the agricultural towns which depended on farm business. Congressman Usher Burdick of North Dakota later declared that the soil bank was "systematically destroying the fabric of North Dakota's small community life." Hundreds of farmers, he said, had rented their farms to the government and moved out of their communities. First, local businessmen felt the pinch, and then the social institutions supported by taxes began to suffer. With a declining number of farmers, Burdick explained, the schools, churches, hospitals, municipal water and sewer systems, and local roads all began to deteriorate. "I was not sent to Congress to legislate the depopulation of North Dakota," he declared.[15]

Challenges to the family farm, a decreasing rural population, and the decline of small towns aroused all of the emotions associated with the agrarian tradition. Most Americans at least gave lip service to the belief that a substantial farm population was important for the political, social, and economic welfare of the nation. They believed that virtue, honesty, diligence, and common sense were more likely to stem from life on the farm than in the city. Secretary Benson expressed the agrarian views with some eloquence in 1960 when he wrote: "We have always had a feeling that there is something basically sound about having a good portion of our people on the land. Country living produces better people. The country is a good place to rear a family. It is a good place to teach the basic virtues that have helped to build this nation. Young people on a farm learn how to work, how to be thrifty and how to do things with their hands. It has given millions of us the finest preparation for life."

According to the Grange, now that the United States was in the atomic age it was more important than ever to support the organization's legislative policies. Those policies reflected "balance and stability, traditional with rural people."[16]

The agrarians saw a direct connection between the nation's farmers and rural institutions and America's greatness. William D. Poe, associate editor of *Progressive Farmer*, told a congressional subcommittee that the small farmer represented "the last great stronghold of freedom in America," and was "the keeper of our Nation's conscience." If the farmer disappeared, he said, "America will lose its earthy flavor, something of its character, something of its independence, and much that has made it great."[17] These sentiments seemed to be as strong or even stronger among those who had never had any direct contact with agriculture as from people with rural backgrounds. President Eisenhower said in 1956 that "no group is more fundamental to our national life than our farmers."

Some writers were lyrical in their praise of agriculture and farm life. The *Daily Oklahoman*, owned by a millionaire urbanite, editorialized on April 20, 1957, that, while farming was "no bed of ease," agriculture provided other compensations. "No man who sees the dawning and hears the whisper of the passing winds and beholds the stately march of the seasons," the editor wrote, "can fail to appreciate that he is among the most fortunate of all God's children. However cruel his lot may be at times and no matter how bitter his disappointments .., he has the realization of knowing that he is living close to the heart of the Infinite." Even popular Washington columnist Stewart Alsop expressed a deep feeling of agricultural fundamentalism and attachment to the land. Writing in the *Saturday Evening Post* in 1956, he told of buying a farm some thirty miles north of Washington, D.C., and asked if "other farm-bred immigrants to the city have completed the same cycle I have." "When you are very young," Alsop wrote, "the country seems a dim and dusty place, and all the world's glory and wonder are in the big city. Then, after some years of walled-in anonymity, you begin to feel a fretful desire for green and growing things . . . and some land, however corned-out and weed-infested, that you can call your own."[18]

It was clear that many Americans still had a strong emotional and sentimental attachment to the land, to farming, and to traditional rural values. The important question, however, was posed by a Kansas farm wife in 1956 when she wrote: "What are we losing along with the little farmer?" "In losing our neighbors," she asked, "have we lost neighborliness, too? In gaining efficiency have we lost the desire or the ability to enjoy a moment of leisure, our kinship with earth, love for the soil? With our mechanical entertainment, do we ever leave TV to watch 'the moon step forth upon the hill?' In gaining security, have we lost vision and strength? With the disappearance of the small farm," she concluded, "America has suffered a serious change."[19] And what about the American character? Wheeler McMillan asked his *Farm Journal* readers in May 1957 if the United States still did not need "a sound back-log of farm-grown, farm-experienced people with farm-grown character?"

The gap between image and reality, the difference between what many people thought conditions should be in rural America and what really existed, was ever widening. Technology, science, and modern management, assisted by government policies, were rapidly creating a new agriculture. Farming could not escape the same trends that were affecting nearly all aspects of the economy. The modern farmer, larger and more efficient, viewed farming more as a business than as a way of life. That phase of American life where millions of small farms dotted the countryside was disappearing. Despite some efforts to help the smaller, poorer farmers to become successful producers, little was accomplished. Economic forces continued to squeeze out the small farmer and favor the bigger and more productive operators. Would the total number of farmers drop to only 500,000 or 600,000, which, according to some USDA officials, would be enough to produce all the food and fiber needed for domestic and foreign markets? Clearly, the trends were in that direction. If farmers did become such a small minority, how would they be able to protect their interests? No answers to these questions were apparent in the 1950s.

Modern Midwest farmer harvesting his corn six rows at a time with farmstead in background. *USDA Photo*

Cutting, thrashing, and sacking wheat in 1902. It took a 33-horse hitch to pull this combine. *USDA Photo*

A self-propelled combine harvesting grain on strip-mined reclamation land in Kentucky, with steam plant cooling towers in the background. *TVA Photo*

The drudgery of working the land with a team of horses or mules has become a thing of the past. *USDA Photo*

This huge rig can plow 60 acres per day; a team of horses can turn about 2 acres. *USDA Photo by Paul Conklin*

Planting directly in corn stubble, this Indiana farmer plants eight rows of corn or soybeans and distributes insecticide and fertilizer in the same operation. *USDA Photo by Joe Branco*

This Iowa hog farm is surrounded by acres and acres of corn. *USDA Photo*

This Mississippi farmer specializes in fresh sweet corn and other vegetables on his 20-acre farm. He sells his produce directly to the consumer at a farmers market. *USDA Photo by David Warren*

Fully mechanized harvesting of potatoes in Michigan. The potatoes are dug from the earth by machine and conveyed to the truck for shipping to the processing plant. *USDA Photo*

Cotton harvesters work their way through a field, each picking as much in an hour as a man could pick in 72 hours. *USDA Photo by Larry Rana*

Workers cut lettuce and trim outer leaves for field packing near Salinas, California. The average yield is 400 cartons per acre. *USDA Photo*

Out to pasture on a midwest dairy farm. *Farmland Industries, Inc.*

Large-scale egg production in Georgia. *College of Agriculture, University of Georgia*

This corn-hog operation in Iowa produces about 2,200 hogs a year—enough to supply the annual pork consumption of about 9,000 people. *USDA Photo*

A cattle feedlot in western Kansas. *Farmland Industries, Inc.*

A feed console operator watches a monitor to see when a truck is in position to receive its load of computer-calculated feed ingredients. She tells the driver by radio which pen in the feedlot gets the feed. *USDA Photo by Michael Lawton*

The desolation of erosion and strip mining. *TVA Photo*

Restoring the land, in Jefferson County, Tennessee. *TVA Photo*

A fungicide is applied to an orange grove in Florida. *USDA Photo*

The circular patterns result from mammoth center-pivot irrigation systems which bring life-giving water to refresh the land. *USDA Photo by Earl Otis*

An abandoned farmhouse in eastern South Dakota starkly reflects the declining number of farms and farmers since 1940. *Photo by Verlyn Larson, Brookings, South Dakota*

American farmers face political realities. A tractorcade protest demonstration in 1979. *USDA Photo*

Farmers Struggle to Hold Their Own

VIII

In 1932 H. L. Mencken, that caustic critic of American life and institutions, declared that one of the best things that could happen to the country would be to get rid of all the farmers. Mencken wrote that such a development would reduce the cost of living, improve politics, and even have a good effect upon religion. The farmer, he continued, was always on the verge of bankruptcy and hated everyone who was "having a better time." According to Mencken, life would be pleasanter without farmers. "This country has been run from the farms long enough," he concluded. By the end of the 1950s some observers were beginning to worry that Mencken's tongue-in-cheek hope might become a reality.

The 1950s was a crucial decade for American farmers. In most areas of the country the transition from the old, less efficient ways of farming to a highly capitalized, more productive type of operation moved ahead rapidly. At the beginning of the decade, for example, there were still a sizable number of two-mule farms in the Southeast, but by 1960 horses and mules had become curiosities for farmers even in areas of less advanced change.

As farmers became an increasingly small minority, only 8.7 percent of the total population in 1960, their political power weakened. For nearly three decades Congress had appropriated huge sums to fund a wide variety of price-support, credit, food, and soil conservation legislation. Expenditures on behalf of farmers continued high in the 1950s, ranging from $4 billion to $6 billion annually, but criticism of farm programs and their cost rose sharply. Indeed, it was the expense of farm programs that aroused much of the opposition to them. The rising consumer movement also brought price-support policies under sharp attack.

Moreover, just as complaints about agriculture's cost to the taxpayers were on the rise, farmers and farm spokesmen were unable to unite behind any particular policies, The farm organizations were divided and party politics could not be kept out of farm policy making. The Democrats, with the support of the Farmers Union, generally pushed for high, fixed price supports, while the American Farm Bureau Federation backed the Republican effort to lower supports and lessen control over farmers. Both parties were sensitive to the political value of farm policies. As House Speaker Sam Rayburn wrote a constituent in 1956, "we are going to try to come up with something with the Democratic brand on it." While there was still considerable bipartisan action on farm legislation in the late 1950s, the group in Congress which had been dubbed the farm bloc began to come unstuck.

An even more important threat to farmer political influence was the demand that state legislatures and the House of Representatives be reapportioned on the basis of population rather than territory. Farmers had been overrepresented for many years, and by the 1950s serious efforts were under way to redress what urbanites considered their lack of fair representation. Thus, while farmers were still a formidable influence in the presidential election of 1956, the kind of power which they had become accustomed to since the 1930s faced serious challenges. A Republican committeeman at Bloomington, Illinois, remarked in the spring of 1956 that "farmers used to run everything in politics here, and now they don't amount to anything—but we've been trying to keep that quiet." About the same time, another observer said that the farmer was a "diminishing political factor even in the farm states."[1]

During the 1950s farm groups began to take a close look at their minority position. In the 1954 legislative report of the Grange, Lloyd G. Halvorson, economist for the organization, discussed this question in some detail. Because of the decreasing farm population, he said, the era when the once-powerful farm bloc could "dominate the legislative forces" was "about over." This situation might not have been so bad, Halvorson explained, if farmers had the economic power to raise their income like labor and business. But that was not the case. Farmers, he wrote, had no choice but to organize more effectively and be sufficiently astute politically so as to "establish by legislation some economic institutions which will help them to maintain a more favorable economic position in our total economy in the years ahead." In other words, farmers should act like the minority they had become. Despite the problems facing farmers, Halvorson believed that "under our political system a well organized minority can have influence far beyond its numbers."[2] That had to be the hope, or farmers would have to resign themselves to an even less favorable position in the economy.

The Farmers Union, the most liberal of the major farm organizations, gave increasing attention to farm politics in the 1950s. M. W. Thatcher, manager of the Farmers Union Grain Terminal Association, told delegates at the 1952 annual convention that political action was extremely important to them. He said that since price supports, credit under the land-bank system, production credit, and the banks for cooperatives were all "given to you by the Congress," "do you agree with me that you live as farmers in a legislative economy?"[3] Farmers Union President James Patton believed that his organization's relations were good with the congressional agricultural committees in the late 1950s, but he blamed the wrongheadedness of Eisenhower and Benson for denying farmers 100 percent of parity and other desired objectives. In any event, Farmers Union officials saw the need for better organization and increased pressure on Congress and on the administration. Unlike the Grange or the Farm Bureau, the Farmers Union urged family farmers to make a coalition with organized workers as a means of winning legislative battles. Patton's friendship with labor unions set him apart from most of the other agricultural leadership in the period.

The American Farm Bureau Federation, the nation's largest and most powerful farm organization, rejected political solutions to farm problems during the 1950s, thereby dividing and weakening agriculture's overall influence in Congress. While supporting some government programs, the Farm Bureau's main thrust under President Charles B. Shuman was, as he put it, to get the government out of farming. He declared in 1956 that "one of the principal causes of the decline in farm net income and of the 25 years of recurring surpluses has been an over-supply of legislative attempts to solve economic troubles by political action."[4] Shuman believed that fixed price supports had helped to create surpluses in commodities for which there was insufficient market demand, and that a freer market would force healthy adjustments by farmers, as well as increase sales overseas. Surpluses in government hands, he argued, acted as a damper or lid on prices.

The Grange stood somewhere between the Farmers Union and the AFBF. It favored a commodity-by-commodity approach, arguing that various farm products required different legislative action. The Grange said that 90 percent of parity had worked well for tobacco but had been a disaster for cotton. For wheat the Grange favored a domestic parity plan that would give wheat producers higher prices for the portion of their crop used at home and a lesser figure for the amount exported.

During the 1930s and 1940s, Ed O'Neal had provided a fairly

united voice for farmers in Washington. His skillful leadership and dominating personality, as well as the needs growing out of depression and war, had produced a remarkable unity among farmers. This, however, evaporated in the 1950s. With different interests, philosophies, and outlooks, Congress found farmer representatives in Washington speaking with many tongues. As a writer in *Capper's Farmer* said in early 1959, congressional farm leaders were "burned up" at the major farm organizations because they could not get together on policy matters. Congressman Cooley, chairman of the House Agriculture Committee, said: "I would like to believe that the leaders of these four organizations [Farmers Union, AFBF, Grange, and National Council of Farmer Cooperatives] could sit down and agree on a major farm program. I am not optimistic." Senator Milton Young of North Dakota referred to "the stubborn and self-righteous attitude of many farm leaders" which had prevented Congress from passing any constructive agricultural legislation in 1959. Senator Joseph Clark of Pennsylvania explained that the most "confounding" problem in getting improved farm legislation was the disagreement among the farm organizations themselves. As if this were not bad enough, there were sharp differences within these organizations.[5] Another major factor contributing to disunity in farm ranks was the close identification of the Farmers Union and the AFBF with the Democratic and Republican parties respectively.

The varied and conflicting interests of farmers were at the root of farm organization differences, making it nearly impossible to get united action behind particular farm policies. This only reflected the situation on the farm, where self-interest prevailed over farm political unity. Poultry and dairy farmers wanted cheap feed, while many cash grain farmers insisted on price supports that kept those feed costs higher. To put farmland into grass as proposed in the soil-bank program, and then permit grazing on those acres, could lead to increasing cattle numbers and hurt cattlemen. In 1956, for example, a Texas hay producer complained to House Speaker Sam Rayburn that paying farmers to grow grass on soil-bank land was unfair to established cattlemen. Why create more competition for cattle raisers, who had not asked for government supports, he asked, in order to pay benefits to wheat and cotton growers?[6]

A nationwide poll of some 4,000 farmers by *Farm Journal* in late 1957 illustrated just how widely farmers disagreed among themselves on agricultural policy. Twenty-seven percent favored more government help, 12 percent thought the current level of government assistance was about right, 11 percent wanted reduced government programs, while 50 percent reported that "government should get clear out of farming." Farmers were a diverse group with such radically different interests that any attempt to lump them together in policy matters was doomed to failure.

The variation of agricultural interests had led to the formation of an increasing number of organizations of farmers producing a particular commodity. These included everything from growers of mushrooms and wheat to those who raised rabbits and cattle. Commodity groups, such as the National Association of Wheat Growers or the National Milk Producers Federation, usually received benefits under the general farm legislation. But like the farmers they represented, they had different interests. By the 1950s the divisions among the major farm organizations and commodity groups had become a matter of deep concern to some agricultural spokesmen.

In August 1957, representatives of thirty-six commodity groups met in Washington to try to work out an agricultural program that all could support. They also invited officials of the major farm organizations. Only the Farm Bureau rejected the invitation. The executive secretary of the National Federation of Grain Cooperatives said that "our job here is to clear the atmosphere and build understanding and find our common cause."[7] After two meetings, however, it was evident that the common concerns were not strong enough to overcome the special interests.

By the late 1950s that bipartisan coalition of senators and congressmen from the Midwest and the South that had been known as the farm bloc was disintegrating. Since 1933 sufficient political power based on an economic community of interest between regions of the country had existed to guarantee agriculture a vast amount of federal aid. This growing lack of unity and agreement surfaced during discussion of a bill to lower price supports for corn in 1957. Midwesterners favored such a move, but southerners opposed handling farm legislation on a single-crop basis. Fearing that a division might threaten price supports for cotton, tobacco, and rice, southerners wanted all major crops included in an omnibus bill. Columnist Ovid A. Martin, who followed agricultural matters closely, reported that the "once-powerful" farm bloc was breaking up under internal stress and strain. Concerned farm leaders and congressmen, he said, longed "for the days when the agricultural forces in Washington were directed by Edward A. O'Neal."[8]

Congressman Thomas G. Abernethy, Democrat from Mississippi, said that the farm bloc was like Humpty Dumpty, "now split down the middle, halved, quartered and splintered." Without a new start and a realignment of farm forces, he said, "a new farm program is hardly more than a lot of talk." Abernethy urged reestablishment of the "once-powerful position of the farm bloc" which had wielded great power. He called for a "love feast" among midwestern and southern congressmen and senators to repair the "cleavage" within the old farm bloc. If constructive legislation were not passed, he concluded, "more small farmers will be gobbled up and our rural population . . . will continue its urban trek."[9] But no "love feast" occurred.

Disagreement by various farm groups, however, was not the major problem facing farmers. Their fundamental weakness came from growing urbanization and population shifts. One political veteran in the nation's capital observed that when he arrived in Washington in the 1930s, some 300 out of 531 congressmen had to pay careful attention to farm needs. By 1957, he said, this number may have been as few as 100. Senator Ralph Yarborough of Texas expressed surprise when he realized how rapidly the farm population in his state had declined. Between 1940 and 1950 the number of his fellow Texans living on farms and ranches dropped from about 33 to 16 percent of the Texas population. Columnist Doris Fleeson wrote that "the breakup of the old farm bloc happened with surprising suddenness because the population shifts took place all at once."[10]

Fleeson called the "utter collapse of the farm bloc" one of the "extraordinary political phenomena of the Eisenhower Administration." Congressmen from the cities who had commonly traded votes with those from rural areas were making fewer deals than formerly. As the high cost of food became an urgent matter among urban residents, city congressmen were less willing to support policies that kept up the prices of raw farm products. Even though studies showed that farmers only received about 40 percent of the food dollar, and that processing, distribution, and other nonfarm costs took the other 60 percent, farmers were usually considered the villains when food costs rose.

By the late 1950s farmers simply did not have the votes or the bargaining power to retain their political position. *Capper's Farmer* editorialized in October 1959 that "city lawmakers are in the saddle, and there isn't much chance the splits in the Farm Bloc will be repaired." The population shifts, according to Fleeson, were "so widespread and so striking [that] many Americans have yet to grasp them." Certainly farmers did not fully realize what was happening to them politically. One politician put it in the frankest terms when he said in 1957: "It sounds brutal to say it, but there are only about five million farmers any more and there are 10 million Negroes: that is why there's no farm legislation before us and we are struggling so hard with civil rights."[11]

The Eisenhower administration, especially Ezra Taft Benson, its main farm spokesman, contributed to dividing the nation's farmers and weakening the voice of agriculture in Washington. While parity was still a kind of unifying symbol, many of the more prosperous farmers agreed with Eisenhower and Benson that it should be achieved in the marketplace rather than by legislation. Raymond Moley, the old New Dealer turned conservative, observed in 1957 that the attitudes of many of the larger and more prosperous farmers were changing. These farmer-businessmen, Moley wrote, had a different outlook than in the past. While they willingly accepted government benefits when offered, an increasing number of them believed that they were efficient enough to make money without so much federal assistance. This is no doubt why *Farm Journal* pollsters found that, of the farmers replying, 50 percent opposed federal involvement in agricultural pricing.

While farmers were declining in numbers, and divided in interest and outlook, they faced another difficult problem by the late 1950s—rising consumer resistance to food prices. Because of American agricultural abundance, people in the United States had enjoyed relatively cheap food. Indeed, over the years they spent a decreasing percentage of their incomes at the grocery store. At the end of World War II, for example, the average family spent 27 percent of its income on food; by 1960 it was only 21 percent. One hour of factory labor in 1960 would buy twice as much food as it would have thirty years earlier. Spending a lesser share of one's income on food left more money in the family budget of city workers for other things that provided a higher standard of living. Millions of American families had come to expect a condition where they spent less on their basic needs and more on their wants.

Since 1933, price supports had been periodically criticized as being a factor in higher food prices. But the complaints never threatened the program. For example, following World War II, as prices stayed high, urban congressmen and senators regularly complained about high food costs and blamed them on federal policy. In October 1949, Congressman Jacob Javits of New York declared that his constituents were opposed to 90 percent of parity because it forced up the price of food. A little later he said that city dwellers believed that consumer prices were being "held up by this parity-price program." Javits added that he believed "in the stabilization of farm prices but it must have a relation to the realities the city dweller faces." Growing opposition to farm programs "by big city districts," Javits said, stemmed mainly from "the high cost of food and the high cost to the people of articles made out of agricultural commodities." Javits was especially critical of higher meat prices. Livestock prices, of course, were not supported by the government, indicating that urban representatives were sensitive to higher food prices from whatever cause.[12]

Despite some criticism of price supports by Javits and others in the late 1940s and early 1950s, a majority of citizens continued to back farm programs. A Gallop poll taken in 1953 showed that of those people who understood the term "farm price supports," 49 percent favored and 45 percent opposed them. Approval was highest among farmers, but 53 percent of the manual laborers also favored price supports. Greatest opposition was found among white-collar workers and the professional and business classes, who, incidentally, lived in cities and were the chief opinion makers.

By the late 1950s a broad attack was being leveled at price supports by urban congressmen, newspaper editors, magazine writers, and other publicists. The campaign was fueled by ever-increasing surpluses, the costs of handling them, and rising food prices. In 1958 total crop production reached an all-time high, and this was matched in 1959. By June 30, 1959, the Commodity Credit Corporation held 1,043 million bushels of corn, 1,146 million bushels of wheat, and 1 million bales of cotton, an amount greatly increased later in the year. The CCC had investments of more than $9 billion in price supports and inventories by 1960. The government held enough wheat to feed the country for two years if not another bushel were grown. In 1958–59, total cost of the farm programs reached $4.6 billion, of which about $1.7 billion represented losses of CCC stocks.[13] It cost hundreds of millions a year just to pay the storage costs.

Secretary Benson's program of lowering price supports and loosening acreage restrictions had failed, if judged on the basis of unneeded and costly farm production. As prices dropped, farmers raised more rather than less in order to have enough production to meet their rising costs and living expenses. For example, in 1958 corn farmers voted against acreage controls for the first time in twenty-five years. They would still receive price supports for the 1959 crop, but at a lower level. Despite huge surpluses on hand and the prospect of lower prices, farmers increased their corn plantings by some 10 million acres in 1959. Prices dropped to around a dollar a bushel, the lowest since before World War II. W.J. Breakenridge and his two sons, who farmed 715 acres of fine land in Iowa, were typical of the larger, more efficient operators who increased their acreage. Breakenridge planted 400 acres of corn in 1959 compared to only 120 the year before. He told a *Life* reporter that farmers should have a fair price for their crop, but that acreage should be controlled. "We can't have price props and unlimited production," Breakenridge said. Yet this farmer, like many others, stepped up his output when he could. Joseph Faivre increased corn acreage on his 640-acre Illinois farm from 52 acres in 1958 to 186 acres in 1959. He said that because support prices were lowered to $1.06 a bushel in 1960 he would plant 287 acres. "With price supports going down," Faivre said, "I have to increase my volume to end up with the same income."[14]

In light of bulging surpluses, congressmen with urban constituencies charged that the price-support programs had been a dismal failure. Congressman Perkins Bass of New Hampshire said that his section of the country was willing to bear some expense to promote the welfare of others, but that the farm program was costing too much. "Most Americans," he said, "would be astonished and deeply shocked at what it cost them to carry on this price support program." What annoyed Bass was that not only did consumers have to pay higher taxes to fund the farm program, but they also had to pay more for food. Looking at his colleagues from the farm states, he asked, "how many more years are you going to demand that we in New England and elsewhere subsidize your agricultural economy?" Congressman Bruce Alger of Dallas expressed the opinion that government incentives were producing the surpluses. This indicated, he said, that "there are now too many farmers on the farms." Jacob Javits, now in the Senate, told his colleagues in May 1959 that consumers were "absolutely opposed to spending more in the blind alley" of "mounting costs and mounting surpluses."[15]

Most of what congressmen and senators said about the folly of the agricultural programs and unacceptable food prices was confined to the dry proceedings of the *Congressional Record*. However, urban newspapers, popular magazines, and even television gave wide publicity to the problem of growing surpluses and discontent with legislation that supported farm prices.

On January 26, 1956, Edward R. Murrow, one of the most influential broadcasters of his time, hosted a CBS TV program on "The Farm Problem: A Crisis in Abundance." The camera showed farm surpluses stored in tents, warehouses, old World War II Liberty ships, and elsewhere. But even government purchases were not saving the small farmer, and the program showed a farm sale near Corning, Iowa, which Murrow said portrayed the demise of the family farm. Secretary Benson fumed at what he considered an unfair presentation, but he could do little to counteract its criticism of current agricultural programs.

Late in 1959 *Life*, one of the nation's most popular and widely read magazines, carried a three-part story on "The Farm Surplus." Calling the farm problem the "most dismaying domestic trouble," the author said that the "government subsidy program" had "sprouted into a national scandal." Earlier in the year, writing from Washington, Columnist Roscoe Drummond said that if something were not done "to halt the Frankensteinian monster of mounting surpluses, aggravated by price supports which pile up bigger surpluses, . . . as sure as a hangover follows a 'lost weekend,' there will be a massive political revolt by the voting consumer." In his book *The Great Farm Problem*, also appearing in 1959, William H. Peterson, a professor of economics at New York University, leveled a spirited attack against current agricultural policies. A writer in the April 1958 *American Mercury* charged that the farm problem was caused by a quarter century of bad legislation, "vote-buying, quack remedies, discarded nostrums . . . rigid support politicians with flexible consciences making hay at the rural voting booths." He called for "de-subsidized, de-controlled, supply-and-demand agriculture." And what about farmers themselves? Of those who answered a *Farm Journal* poll in 1959, approximately 55 percent said they wanted the federal government "clear out" of farming. This was up 5 percent from only two years earlier.[16]

During congressional discussion of a new farm bill in 1958, the relationship between farm policies and the cost of living received close attention. Urban critics of the measure opposed it because they believed it would raise food prices. One of the main reasons the bill was defeated was that city Democrats deserted farm-state members of their party on this issue. They referred to the bill as a "bread tax," a "milk tax," and a "tax on consumers." There were additional reasons why the measure did not pass, including partisan politics, but as a writer for the *Farm Journal* commented, the debate over the agricultural bill in 1958 made it clear that in the future consumers would have a major influence on farm legislation. More and more urban congressmen and senators were confirming this view. When Senator Joseph Clark of Pennsylvania was asked how city lawmakers viewed farm legislation, he replied that if parity prices meant "a slight increase in retail food prices, no city lawmaker could justifiably vote for it." There would be serious objections from consumers, he continued, "if we adopted any farm program which would raise retail prices." While this was an exaggeration and some urban congressmen continued to vote for price support legislation, there was growing opposition to such legislation.[17]

Indeed, the rising consumer movement brought a new element into the question of government policies to support farm income, and placed farmers on the defensive. Up to the 1950s supporters of farm legislation paid little attention to how those policies might affect other groups in society. Farm policy was for farmers. Their spokesmen were the farm organizations, the agricultural committees in Congress, the Secretary of Agriculture, and the USDA. Broader benefits stemming from the distribution and price of food were more or less incidental to the larger question of helping producers. The distribution of surplus food in the 1930s, the food stamp plan begun experimentally in 1938, the school lunch program, and various schemes to ship food to needy foreign countries came about mainly to help get rid of surpluses. They represented fallout benefits to consumers both at home and abroad from the main agricultural programs, which were really designed to help farmers by propping up prices.

But consumers were getting more demanding. In some cases they insisted on having something to say about farm policy. As Javits said in 1950, agricultural policy was "not just a problem for farmers. It is time," he said, that "we had a department of consumers in the Federal establishment and consumers legislative committees in both Houses of the Congress." To farmers such a proposal could mean only one thing—nonfarmers would exert increasing influence on government policies relating to agriculture.[18]

Commercial farmers deeply resented the idea that consumer interests might intrude into agricultural policy making. They believed that consumers were only interested in low farm prices and cheap food, a policy that would hurt farmers. They also resented the assumption by urban spokesmen that city residents somehow deserved low food prices. Farmers saw their city cousins spending a smaller and smaller percentage of their incomes on food, and asked bluntly why farmers should subsidize urbanites through low, unprofitable farm prices. From the viewpoint of many farmers, the main benefits from their hard work and high productivity were going to the rest of society. It was frightening to hear Senator Javits and others talk about a department of consumers. With their declining numbers, farmers already felt as though they had little enough power to protect their interests without another special interest group opposing them. No doubt many agreed with the York County, Pennsylvania, farmer who said in 1959: "It seems that each year agriculture loses position and importance."

Making matters worse from the perspective of farmers was the continuing talk about implementing programs that would further reduce the farm population. On one occasion Javits suggested paying farmers while they were being retrained for other employment. He said it was completely "archaic" that "we can have a yeomanry in the United States." "Sentimentally," he said, Americans would like to keep this idea alive, but it had no meaning. Erwin D. Canham, president of the United States Chamber of Commerce, echoed this view. He told a Kansas City audience in October 1959 that there were too many farms. "It is true that the farm environment is spiritually and morally valuable to our national society, but the farmer," Canham said, "is not a museum piece. He, too, must go along with the changes which are inevitable in our type of economic order. We may love the self-sufficient family farm, but we cannot endow it in perpetuity. It must rise or fall on its own." Writing in the *Washington Post* in January 1957, J. A. Livingston said that the principal function of government farm programs should be to give poor farm families training and guidance for urban living, and assist them "to abandon mean, unrewarding acres." The main task, he continued, was "to take people out of cultivation." Such statements aroused latent agricultural fundamentalism, and convinced farmers that an effort was being made to completely destroy farmer influence in the nation.[19]

Farmers reacted even more strongly against charges that they were getting rich off government programs. In 1955 a story in the *New York Times* stated that "the ordinary Iowa farmer . . . has a minimum of two cars and they are usually brand new Buicks or Oldsmobiles or Cadillacs." Actually, only one-half of one percent of Iowa farmers drove Cadillacs in the mid-1950s. Farmers knew that such stories were sheer nonsense, but many city people confused myth and reality. The fact was that even in Iowa more and more farmers, or members of their families, were taking nonfarm jobs to maintain a modest standard of living. One farmer near Newton said in 1958 that "the standard of living we have we wouldn't have if we weren't working off the farm. We have good machinery, laborsavers, a new automobile. Couldn't do it if we didn't have other income."[20]

How could farm spokesmen counteract the perception that farmers were greedy claimants of government bounty who drove Cadillacs and were responsible for rising food costs? The gut reaction of farmers was to resort to some kind of public relations program that, from their viewpoint, would set the record straight. This approach, however, faced serious difficulties. The main problem for farmers was that the principal organs of public opinion—newspapers, magazines, radio, and TV—were controlled by urban consumers. It was the *New York Times* that reported Iowa farmers driving big and expensive automobiles, and *Life* which called farm price supports a scandal. The means that farm organizations had to advertise their side of any issue was limited to the writings of some excellent farm editors in such papers as the *Des Moines Register*, in the farm magazines, most of which had only state or regional coverage, and in publications of the national farm organizations, cooperatives, and commodity groups. The farm newspapers and magazines, however, circulated mainly to farmers. The nation's farm press was in a very real sense preaching to the converted. *Farm Journal*, the only truly national farm periodical, had a circulation of some 3.3 million in 1960, but it, too, went principally to rural residents. Moreover, in the 1950s most farm groups had small budgets for publicity and advertising campaigns. To make matters even worse, leadership in the Farmers Union and the Farm Bureau was so occupied with pushing their favorite programs in Washington that there was little time or energy left to deal with the bigger problem of the farmer's image.

For farmers, then, setting the record straight, as they put it, was a difficult task. Nevertheless, some farm spokesmen tried their best. *Capper's Farmer*, for example, included in its December 1958 issue an insert entitled "What City People Should Know About Farmers." Admitting that "there's no real spokesman for farmers; no one to filter the facts from a fog of headlines and oratory," the writer went on in text and pictures to defend price supports and to show that rather than being responsible for higher food prices farmers had actually subsidized consumers between 1952 and 1957. "Farmers are a dwindling minority in a growing America," said the writer, who urged "ALL of us" to work harder to tell everyone how vital agriculture was to national security and the country's economic well-being. The *Progressive Farmer*, edited by that venerable agricultural spokesman Clarence Poe, headlined his story "Let's Tell the World These Farm Facts." Poe pointed to high costs of operation and to low net income to farmers. On the other hand, Poe wrote, consumers were getting relatively cheap food, while labor and industry enjoyed numerous governmental benefits that seemed to help keep them in a much better position than farmers. Other farm papers echoed these sentiments.

Farm-state congressmen and senators could be counted on to advance the farmer point of view in congressional debate, committee hearings, and in public speeches. A subcommittee of the House Committee on Agriculture held special hearings on food price trends in 1957 to learn just who or what was responsible for higher grocery prices. Chairman Cooley said that housewives went into markets and paid 10 cents for a tomato and 13 cents for an apple and jumped to the conclusion that farmers were getting rich. Somehow that assumption needed to be corrected. In testifying before the subcommittee, representatives of the major farm organizations did not agree on the causes of higher living costs, but they did concur that farmers were not at fault. When urban congressmen complained of rising food prices, those representing agriculture jumped to the farmers' defense. Senator Hubert Humphrey reminded his colleagues during such a debate in 1959 that even if farmers gave wheat away a 19.3-cent loaf of bread would only save consumers 2.4 cents. Indeed, bread prices sometimes rose when the market for wheat dropped. This illustrated a vexing problem for farmers, who saw themselves as the victims of false and unfair criticism. Governor Orville L. Freeman of Minnesota, soon to become Secretary of Agriculture, told a congressional committee in 1960 that he was "very weary . . . hearing about how farmers were subsidized." He insisted that the farmer had "been subsidizing the consumer." In 1962 President Kennedy declared that the great gains in agricultural productivity had been "passed on to consumers."[21]

However hard they tried, farm spokesmen had little success in converting city residents to their position. In the battle for public opinion, farmers were at a clear disadvantage. Indeed, it was naive to believe that they could solve their problems through any kind of public relations drive. They had to rely more on political influence in Congress to protect their interests than the goodwill of city residents.

The growing shift in political power toward the cities had been a matter of concern among farm spokesmen for several years, but the one man, one vote decision by the United States Supreme Court in 1964 struck agriculture like a thunderbolt. It seems clear enough that farmers and their state and national representatives were only faintly aware of the drive that began in the 1950s to reduce agriculture's political power. Traditionally, most state constitutions had given rural residents a disproportionate number of seats in the state legislatures. Congressional districts also failed to reflect urban growth. On a population basis farm votes carried more weight than those in the cities. However, representation in terms of geography, which favored farmers, came under sharp attack in the mid-1950s. Despite a Supreme Court case in 1946 which rejected a plea for reapportionment, urban citizens demanded redistricting on the basis of population, rather than land area.

In 1956 liberal columnist Richard Strout wrote in the *New Republic* that conservative rural legislators were blocking reform, and that representatives in Congress were responsible for the expensive agricultural programs and high grocery prices. Senator John F. Kennedy joined the liberal cause and claimed that unequal apportionment was the root of growing urban problems. Cities could not get what they needed, Kennedy added, because "the urban majority is politically a minority, and the rural minority dominates the polls." Other writers accused farmers of holding the reins of power and using that power to fatten themselves at public expense. "It took the farmers a long time to pry open the doors of the federal treasury," declared one observer, "but since they succeeded the bounty has been unceasing."[22]

By the late 1950s a reapportionment battle was quietly shaping up in several states. Citizens in Denver, for example, initiated a suit in 1956 to compel a change in representation between urban and rural residents in Colorado. Observing this development, a writer for the *Farm Journal* said that "basically the question is whether farmers will give up some of their political strength to the cities." In 1959 a suit was filed in the federal court in Tennessee to force reapportionment of the legislature as called for in the state constitution. Failing to win there, the case was appealed to the United States Supreme Court, which held in *Baker* vs. *Carr* in 1962 that state legislative apportionment was subject to constitutional review. This opened the way for other suits. In 1963 the Supreme Court outlawed Georgia's county unit system that had given rural areas preponderant power in state elections. The next year the high court delivered two sweeping decisions which held that congressional districts, and both houses of the state legislatures, must be apportioned according to population, not geographic area. This was the one man, one vote doctrine.

The Supreme Court's 1964 decision took most farmers and their representatives by surprise. They somehow seemed to assume that one branch of the state legislature would always be based on land area. When the decision was announced, most farm spokesmen quickly came to the defense of the current political arrangements. Claude W. Gifford, economics editor of the *Farm Journal*, wrote that the old system permitted "representatives of farm sections to better serve widely scattered farm people and their economic and minority interests." Charles Shuman, president of the AFBF, said that the principle of one house based on population and one on area assured the protection of minority interests, and contributed to a "check-and-balance system that has worked well for 188 years." But, as Gifford said, the ruling tossed "into the ash can one of the basic, time-honored cornerstones of our system of American representative government." Speaking at the annual meeting of the Consumers Cooperative Association, one of the nation's largest farmer cooperatives, in December 1964, CCA President Homer Young said that agriculture "must face the reality of a dwindling minority voice in government."[23]

But even on this issue farmers and farm organizations were not united. James Patton, president of the Farmers Union, believed that the decision was just and entirely consistent with democratic principles. In a poll taken by *Wallace's Farmer* in February 1964, farmers themselves expressed indecision over reapportionment and its significance for agriculture. In general, however, farm interests recognized that reapportionment, especially realignment of congressional districts on the basis of population, would weaken the political power of agriculture. The one man, one vote principle could only contribute further to the declining influence of farmers. "This fading importance of rural America is brought sharply into focus in the recent U.S. Supreme Court ruling," wrote an editor of the *Kansas Farmer*. He added that "it is going to take thoughtful legislators and sound legislation to protect farms and ranches on both the state and national level in this country."

What, if anything, could be done? Several bills were introduced to limit the Supreme Court's jurisdiction in state apportionment, or to guarantee states the right to decide the matter without interference. The main effort to retain greater farm influence, however, was headed by Senator Everett Dirksen of Illinois, who led a fight in Congress to pass a constitutional amendment permitting the states to apportion one house of the legislature on the basis of territory rather than population. Herschel Newsom, Master of the Grange, and Shuman of the AFBF, both supported the Dirksen effort. They were supported by some of the commodity groups and farmer cooperatives. Considering the realities of America's population in the early 1960s, any attempt to amend the Constitution to preserve disproportionate representation was unrealistic and futile. The Dirksen amendment did not get the necessary two-thirds vote in the Senate. Farmers simply had to accept and learn to live with their dwindling political power. Many rural residents agreed with Senator George Aiken of Vermont, who said in late 1964 that within a few years control of the nation would pass into the hands of "big city political machines." Aiken also predicted that after the census of 1970 agriculture would lose "much of the protection and services it now received under federal law." In 1890 A. B. Hart, the famous Harvard historian, had written that the growing process of urbanization would eventually result in the cities being "the great controlling force in the affairs of the nation." By the 1960s, that time had come.

The decline of rural political power caused some farmers and their leaders to consider attacking their problems in another way. If it were true, as an Iowa country editor wrote in November 1958, that state and national politicians were "no longer afraid of the farm vote," economic organization might be the only feasible way for farmers to protect their interests. During the 1950s and 1960s, a growing number of farmers were willing to try this approach.

Bargaining Power for Farmers

IX

FARMERS UNDERSTOOD THEIR BASIC PROBLEM clearly enough. They simply lacked the organization and bargaining power enjoyed by business and labor. Whether a farmer was among the lower- or middle-income groups, or was one of those who sold $40,000 or more worth of products annually, his problem was essentially the same. Unlike other businessmen, farmers did not set the prices either of what they bought or of what they sold. When they reached the markets for cotton, wheat, or hogs, farmers asked "What will you give me?" and when they bought seed, machinery, fertilizer, or other production items, the question was "What must I pay?" Except in times of unusual demand for agricultural products, such as during World War II, farmers were largely at the mercy of buyers and sellers.

Why could not farmers operate like other businessmen and set the price of their commodities? Instead of saying to buyers "What will you give me?" why should farmers not say "This is the price we must have to pay our expenses and make a profit"? The obvious answer was that they did not control their production or the flow of commodities to market. Farmers consistently produced surpluses, which depressed prices. Moreover, with some exceptions, production and marketing of agricultural commodities were done on an individual basis, without plan or organization based on what other farmers were doing. The federal government had developed a degree of united action by encouraging marketing agreements for some commodities, as well as by providing price supports. But many farmers resented government controls and were looking for other ways to increase their influence in the market. By the 1950s and 1960s there were increasing demands for bargaining power for farmers. This meant some kind of united action. As Carroll P. Streeter wrote in the *Farm Journal* early in 1958, "They've got to band together into powerful groups to do what they can never hope to do alone."

The idea of farmers working together to increase their bargaining power was not new. This was the impetus behind the cooperative movement that had been growing since the 1870s. The Grange and the Farmers Union had both supported the formation of farmer cooperatives before World War I. When the American Farm Bureau Federation was organized in 1919 it, too, backed the cooperative movement. By the 1920s there were hundreds of producer and consumer cooperatives throughout rural America. The marketing cooperatives did not pay farmers more for their product than the going price. Rather, they emphasized orderly marketing and used whatever influence they could exert to get higher prices. It was also common for some commodity cooperatives to pay farmers only part of the purchase price at the time of delivery and the remainder after the product was sold—for a better price, it was hoped, than that offered at the time of original sale. The consumer cooperatives operated on the English Rochdale system. That is, they charged their farm customers regular competitive prices and at the end of the year returned a portion of the profits, or net savings, to members on the basis of the amount of business each had done with the association.

During the early 1920s, the farmer cooperative movement spread rapidly. Although ordinary dirt farmers were less enthusiastic, some promoters believed commodity marketing cooperatives would be a panacea for producers. Between 1920 and 1925 commodity cooperative marketing associations developed for tobacco, cotton, wheat, peanuts, and other crops, as well as dairy products. Some farmers, especially tobacco growers, were subjected to high-pressure tactics to get them to sign up and deliver their crop through the cooperative. In 1923 tobacco cooperatives handled about 44 percent of all tobacco sold. By 1924 and 1925, however, most of the new marketing cooperatives were in trouble. Many farmers opposed the contracts they were required to sign in order to sell through the cooperatives, and they objected to not receiving full payment at the time of sale. The cooperatives also experienced management problems, and they lacked capital to carry on large-scale marketing activities. By the mid-1920s, it was evident that, as one economist said, cooperatives could not fix prices, eliminate middlemen, or greatly cut marketing costs. After 1925 there was a shift away from the idea that cooperatives might control farm prices and to the objective of increasing marketing efficiency.

While the tobacco, cotton, grain, and livestock marketing cooperatives failed to affect prices for their members in the 1920s, some of the fruit, vegetable, and dairy associations had considerable success. For example, by 1929 the California Fruit Growers' Exchange controlled some 74 percent of the citrus shipments from California and Arizona. The Dairymen's League Cooperative Association represented more than 43,000 dairymen in New York and helped to stabilize milk prices for producers. Following some hard times during the 1930s, and passage of a federal milk marketing order in 1937, the Dairymen's League and other cooperatives were able to stabilize milk prices in the entire New York milk shed. However, without federal price supports for fluid milk, and the government purchase, storage, and distribution of other dairy products, the cooperatives would not have been able to control marketing for dairymen. During the 1920s and 1930s, farm marketing cooperatives had a checkered record. Some advances were made, but with the exception of successes in dairying and certain specialty crops, these associations were unable to gain enough bargaining power to influence agricultural prices effectively.[1]

Farm welfare did not, of course, depend on the price of agricultural products alone. As nonfarm inputs—gasoline, chemicals, nutritionally balanced feeds, machinery, and other commodities—became more important in farm operations, the cost of these items directly affected profit margins. If farmers could establish successful cooperatives to handle these supplies, they might be able to reduce the cost-price squeeze. During the 1920s, an increasing number of consumer cooperatives were set up among farmers in the Midwest and elsewhere. For example, the first farmer cooperative association formed to handle petroleum products exclusively was the Cottonwood Oil Company of Cottonwood, Minnesota, organized in July 1921. These farmer cooperatives bought supplies at wholesale, usually from noncooperative sources, and resold them at retail. The consumer cooperative associations hoped to provide farmers with good products at reasonable prices, and at the same time pay cash refunds to members.

Perhaps this development can be best explained by looking at the development of Farmland Industries, Inc., a highly successful farmer cooperative. Organized by Howard A. Cowden in Kansas City, Missouri, as the Union Oil Company in 1929, this wholesale cooperative set out to provide petroleum products to member associations. Initially, only six local associations affiliated with the Union Oil Company, but within a few years scores of cooperatives joined up by taking stock in the wholesale.

While furnishing gasoline, oil, and grease on a cooperative basis to farm customers might have been of some help, Cowden had much bigger plans. He believed that farmers must not only distribute products through their cooperatives, but that they should also produce them in their own plants. Impressed by integration in American business, Cowden envisioned the day when farmers would gain control over all the processes of production and distribution, from raw materials to finished products. The most significant savings to farmers, he argued, would come in manufacturing, not in distribution. So long as farmer cooperatives had to rely on Standard Oil or some other large company for their source of petroleum supplies, savings to farmers would be small. Therefore, he worked toward obtaining farmer-owned oil wells, refineries, a grease plant, minerals, fertilizer plants, feed mills, and other facilities that could produce needed supplies for modern farmers.

By the mid-1950s, after years of struggle, the Consumer Cooperative Association, as it was then named, had spread to seventeen midwestern states. Some 400,000 farmers in more than 1,600 cooperative associations were doing in excess of $90 million worth of business annually with their own company, CCA. Not only did farm members receive high-quality petroleum products, feeds, fertilizer, paint, and other products, they also enjoyed hundreds of thousands of dollars in refunds. Farmers were proud of the three oil refineries, the first fertilizer plant built by a cooperative (in 1954), and other facilities which they owned and operated through membership in their local cooperative associations.

Other regional cooperatives were also growing and serving hundreds of local farmer associations. These included Gold Kist, based in Atlanta and the largest farmer cooperative in the Southeast, the Illinois Farm Supply Company, the Farmers Union Central Exchange, and numerous others. By 1955 there were 7.7 million farmer memberships in nearly 10,000 marketing and supply cooperatives.[2]

But after years of effort, cooperatives had not significantly increased the bargaining power of farmers. In the 1950s costs kept going up while many agricultural prices dropped, creating the familiar cost-price squeeze. Farm cooperative leaders kept talking about expanding the size and strength of the associations. Marvin J. Briggs, retired general manager of the Indiana Farm Bureau Cooperative Association, declared in 1957 that "to attain dignity, equality and parity in our present economic structure, the farmer must answer big business with big cooperatives."[3] Farmer associations did grow significantly, and they served more members and did more business than ever in the 1950s and 1960s. They also provided reliable services to farmers. With few exceptions, however, they failed to provide countervailing power in the marketplace. Why?

Only a few cooperatives, such as those organized to handle fruit, vegetables, nuts, and dairy products, met the requirements for successful bargaining power. To be effective, cooperatives had to regulate output or control the distribution of the commodity after production. In the case of milk, the cooperatives controlled the disposition of the product. They bargained with the processors over the price to be paid for grade A milk and then diverted any surplus into the so-called secondary market, where it was processed into cheese, dry milk, and other products. Without this secondary market for surplus milk, the cooperatives would not have been able to negotiate satisfactory contracts with distributors of fluid milk. Furthermore, government price supports placed a floor under milk prices. In the case of fresh fruit, the cooperatives sometimes destroyed surpluses to keep them from depressing the market. Thus, in order for farmers to bargain effectively, their agency had to control a sufficient volume of a product wanted by processors, it had to have disciplinary power over members, it needed to be able to inflict losses on buyers, and as a last resort member farmers had to be willing to withhold production from market. Most cooperatives did not have this kind of control.[4] Many farmers, however, had not abandoned the principle of increasing their market bargaining power.

On a September evening in 1955 a group of corn-hog farmers met in the basement of the Okey-Vernon National Bank in Corning, Iowa. Led by Wayne Jackson, a farmer, and Jay Loghry, a feed salesman, the group discussed what they might do to improve their economic condition. They agreed that somehow they must strengthen the farmers' position in the marketplace if they were ever to overcome the cost-price squeeze. The outcome of this meeting was the National Farmers Organization. Oren Lee Staley was elected president at a national convention held the following December at the National Guard Armory in Corning. Staley, a large, round-faced farmer who raised corn and livestock on his 400-acre farm near Rea in northwest Missouri, believed with evangelical fervor that only through organization and collective bargaining could farmers improve their position in an economy where business and labor were highly organized. The NFO was a grass-roots movement composed of farmers who were fighting mad because they were not experiencing the prosperity enjoyed by other elements in the economy. These were not down-and-out farmers on their last economic legs.

The NFO's plan was to organize farmers, then seek master contracts with processors and buyers of farm commodities at prices based on the cost of production, plus a profit. If processors refused to enter into collective bargaining agreements, then the NFO would stage a "holding action" in an effort to force buyers into an agreement. Without the will and ability to withhold commodities from processors, Staley said, there could be "no real bargaining." These were the steps that the NFO believed were necessary to build for farmers what Staley called "counteractant economic forces."

The NFO established headquarters in Corning, and sent scores of organizers hurrying about midwestern communities to sign up members. The movement caught on quickly, especially among hog raisers, whose income had taken a sharp slump. In talking to producers about increasing their bargaining power in the marketplace, the NFO struck a responsive chord among many younger farmers who were burdened with high capital costs and operating expenses, and who were trying to maintain just a modest standard of living. A Henry County, Illinois, farmer who had begun farming in 1951 said that he had been forced to sell his hogs before they were ready for market because he was out of feed and did not have money to buy more. He was cutting his expenses every way he could. A young Iowan said that his electric bill was $30 a month, but he could not cut that cost. If he tried to make his wife live like he did before the days of electricity in order to save money, he said, "I'd have to tie her to a post first."

While the NFO never revealed the number of members it recruited, during the next few years it signed up thousands of farmers. The most common membership figures given varied from 125,000 to 180,000. Most members were in the Midwest. The rapid growth of the NFO in strong Farm Bureau and Farmers Union country indicated that many farmers did not believe that the older organizations were really coming to grips with problems down on the farm.

At the same time, the NFO created tensions in some communities as organizers placed heavy pressure on farmers to join. If the NFO were ever to gain a position where it could force processors into collective bargaining agreements, it was essential that it control a substantial portion of the commodity. NFO leaders usually mentioned 60 percent. To do this the organization needed as many members as possible. Besides resenting some high-pressure tactics, many of the more conservative farmers rejected the NFO's contacts with organized labor and the talk of a farm strike. In any event, within six months of the organizational meeting in December 1955, the NFO was receiving national attention. When *Time* ran a story on farm leadership on May 7, 1956, Staley's picture appeared beside those of Patton, Shuman, and Newsom. It was no longer just the big three; the NFO had changed agricultural representation to "The Farmers Four Voices." Institutions and organizations often rise up to meet a special need in a particular time and place. The NFO was such an organization, at least from the viewpoint of between 100,000 and 200,000 American farmers.

Although holding actions had been tried in 1959 and 1961, the first major test of NFO strength did not come until 1962, when low livestock prices triggered farmer action. On August 28, about 20,000 farmers gathered at the Veterans Memorial Auditorium in Des Moines to hear Oren Staley call for a holding action which would keep farm products off the market until higher prices had been negotiated and contracts signed with packers, food processors, and other buyers of agricultural commodities. Staley, an old-style evangelistic orator, played to the emotions of a packed audience when he shouted "Are you with us?" and the thunderous response came back—"Yes!" Staley said that farmers were "the most underpaid group in America," and that it was time for farmers to get some control over their economic welfare. The convention endorsed withholding farm products from the market until grain and livestock processors paid $22.75 a hundred pounds for hogs, $32.45 for choice cattle, $1.49 a bushel for corn, and $2.56 for soybeans. These prices compared with current market prices of $19.00, $28.00, $1.10, and $2.30 for hogs, cattle, corn, and soybeans respectively. The holding action, or strike as it was sometimes called, was to begin at midnight, Friday, August 31. Staley said that "we want to get ourselves in a position where we would be as effective as industry in pricing and effective as labor in bargaining."[5]

The NFO's main effort was to convince farmers not to deliver hogs and cattle. Members called "minutemen" or "checkers" set up checkpoints to observe the holding action's effectiveness. Staley explained that these farmers were not stopping trucks, but were "just keeping their eyes open to see who brings in anything." Some minutemen, however, did not hesitate to intimidate producers who did not cooperate with the holding action. During the first and second weeks of September there was a rather sharp decline in receipts of hogs and cattle at the twelve major midwestern markets. On September 5 the USDA estimated cattle receipts at 42,400 compared to 50,024 a week earlier, and 51,132 a year before. Hog receipts were down about 20 percent. As deliveries dropped, prices rose slightly. Hogs reached $20 a hundred pounds on September 5. By September 6 packing plants began to lay off employees, and wholesale meat prices rose several cents a pound in Chicago and New York.

Meanwhile, violence flared up in some communities as strikers tried to discourage anyone from sending hogs or cattle to market. Two farmers from Savannah, Missouri, said that their truck had been shot at, while others reported broken windshields, pierced tires, cut fences, a burned barn, and other damage. Staley insisted that NFO members were not responsible for such activities, but few outside the organization believed his disavowals.

When nearly a month had passed and the NFO had been unable to force processors to sign contracts for livestock, support for the holding action dwindled. Faced with failure, Staley called a "recess" in the holding action on October 2 in order, he said, to give the NFO a chance to consolidate its gains and to prepare to exert more pressure later. Farmers rushed livestock to market, some of which had become heavier than demanded for top quality, and prices slumped to near or below where they were before the holding action began.

In August 1964, mainly in response to low hog prices, the NFO launched another holding action. The goals were the same as two years earlier—keep livestock from market, and force processors to sign contracts that would give farmers higher prices. Again there was strong pressure throughout the Midwest to keep farmers from delivering their livestock to market. In late August and September, cases of slashed tires, shootings, and other acts of sporadic violence occurred. The most serious incident happened September 9, when some five hundred farmers blocked hogs and cattle from reaching the Cooperative Livestock Sales Exchange at Bonduel, Wisconsin. Several trucks were turned back. One driver, however, refused to retreat and as his truck moved forward two of the pickets fell under the wheels and were killed. The police saved the driver from the angry crowd. Despite picketing and pressure on farmers not to sell livestock, this strike by the NFO soon ended.

Failure of these early holding actions did not deter angry farmers from further efforts to gain control over their prices. In March 1967, following a period of very unprofitable dairy operations, the NFO called for a strike against the delivery of milk in twenty-five states to force processors to increase milk prices. Secretary of Agriculture Orville Freeman called the milk strike a symptom of a "chronic [economic] sore." In order to shut off deliveries, farmers attracted widespread attention when they poured thousands of gallons of milk in ditches and sewers. The evening telecasts horrified some urban housewives as they saw the white liquid running down the street curbs. One city housewife said that, although she had been sympathetic to farmers earlier, "now I wouldn't pay them 2¢ more for a tubful if I could help it." One widely distributed picture showed a Dane County, Wisconsin, dairyman dumping a day's production—about $700 worth—from 450 cows. When critics argued that farmers would not benefit from such actions, another Wisconsin farmer said, "I'd rather be hurt now than slowly strangled" from low prices.

Despite a heroic effort, the NFO was unable to shut off milk supplies to the point where real shortages developed in the major urban markets. As NFO supporters tried to shut down deliveries, some violence flared up between NFO members and noncooperating farmers who disagreed with such tactics. Although the strike was already weakening by the close of the second week, the end came on March 29 when the Department of Justice obtained a temporary injunction against certain NFO actions. As it turned out, about all farmers had done was to dramatize their predicament of low prices and high costs. The editor of the *Farm Journal* wrote that "as a propaganda maneuver it did some good." Unfortunately, the strike divided communities and aroused animosities among neighbors so that farm unity was reduced even further.[6]

Who were these militant farmers who were willing to block the entrance to processing plants, dump milk on the ground, and intimidate neighbors who might want to market livestock in spite of NFO pleas? Critics of the organization commonly charged that most NFO members were the poorer and less successful farm operators. It was said that since they were not good managers or very efficient, they turned to strong-arm tactics. This image of NFO members was found to be inaccurate by two rural sociologists who surveyed about 8,700 midwestern farmers, of whom some 1,200 were NFO members. These scholars found that the income of NFO farmers was as good as farmers who were not in the organization. However, the survey did find that NFO members were younger, more heavily in debt, had worked more away from the farm, and had had greater contacts with labor unions. Also, they seemed to be less satisfied with the economic disparity of farm incomes. Their greater economic hopes and expectations seem to have been a major factor in their militancy.

When farmers failed to receive what they considered fair prices, they continued to dramatize their condition in the press and on TV. In the spring of 1968, NFO members killed several hundred calves and mature cattle, and threatened to destroy 12 million pounds of meat if packers did not negotiate prices and sign contracts. But again direct action did not produce concrete results.

Dairymen and livestock producers were not alone in wanting to call their plight to public attention. In July 1969, a group of mainly grain farmers left Redmon, Illinois, on a "tractor march" to Washington, Led by Cleo A. Duzan of Oakland, Illinois, 150 farmers drove their tractors toward the nation's capitol. "We're off to break the economic chains that bind farmers," Duzan announced. Called the farmers' survival drive, this was the country's first tractorcade, and the organizers hoped to arouse Congress and the President to the poor economic position of grain farmers. Arriving on July 28, in Beltsville, Maryland, near Washington, the tractor drivers were joined by several hundred other farmers who had arrived in Washington by pickup and plane. Scores of irate farmers jammed the House Agriculture Committee room and urged restricted production and full parity prices. One young farmer said, "I don't think the farmer is getting his fair share of the economy. The time for complacency is past." Farmers had gained national attention, but Congress ignored their special demands.

The strikes, destruction of milk and livestock, and a tractor march to Washington all reflected the increased willingness of at least some farmers to use direct action to seek redress of their grievances. A few of these actions were mainly media events, not unlike those staged by other protest groups in the 1960s. It was interesting to note that some of the same farmers who had been highly critical of student protests and civil rights marches were seen trying to block trucks from delivering livestock and were riding tractors to Washington. There was certainly no revolution in rural America, but it was clear that the spirit of Daniel Shays was not entirely dead.

The NFO continued to have a strong appeal to many farmers because it expressed more clearly than other farm organizations one of the farmer's basic difficulties—lack of control over the prices paid for his products. The NFO knew the problem, but it did not have a workable solution. Farmers lacked the essential elements necessary for successful collective bargaining and price setting. No groups producing grain or livestock had control over sufficient volume of a product to force processors to meet their demands. Buyers could find alternate supplies. Moreover, farmers had neither the cohesion and unity to be effective bargainers nor the needed disciplinary power over their members. In this respect farmers differed from organized workers. For example, during the 1962 holding action two young Iowa farmers said they liked the approach. One had held his hogs even though he was not an NFO member. An older farmer, however, opposed the holding action and said: "I like a free market."

But the main weakness of a farm strike is that livestock and perishable commodities cannot be held off the market very long. Milk cannot be stored, and if farmers do not sell it they must pour it out. Fruits and vegetables spoil. Hogs and cattle gain weight while they are held, costing the owner for feed, adding to the surplus of meat, and losing the top market. When hogs, for example, reach 200 to 240 pounds, they bring top price. Less is paid for heavier weights. As the *Farm Journal* editor said, "packers can outwait farmers." Even in the case of storable commodities, such as grain, most farmers were not in a financial position to withhold them from market for an extended period. They had bills to pay and operating expenses to meet. But apart from the economic problems connected with any withholding action, the majority of farmers were not willing to accept the discipline and adopt the tactics necessary to improve their situation through collective bargaining. It smacked too much of the tactics of organized labor, whose actions were widely criticized by many farmers.

Secretary Freeman and some members of Congress talked about giving farmers special collective bargaining rights, or "muscle in the marketplace," as the Secretary called it. Freeman told delegates to the National Farmers Union convention in March 1967 that "farmers have power if they act together." That was the trouble, of course—farmers would not work together. They were, as one writer put it, "cussedly independent as always." In 1967 Congress passed a watered-down Agricultural Fair Practices Act, purportedly to increase farmers' marketing power, but it was of little help. The next year Senator Walter F. Mondale introduced a bill which would have provided for a National Agricultural Relations Board, comparable to the National Labor Relations Board, but the bill turned out to contain a hodgepodge of provisions to which neither farmers nor Congress could agree.

The early history of the National Farmers Organization reflected deep frustration and even exasperation among a growing number of farmers. As they viewed it, everything was working against them. With their numbers declining and economic pressures mounting, they sought to protect themselves through tighter and more effective organization. While the leaders would not admit it, by the 1960s the NFO had done little for farmers except to publicize their problems. The older farm organizations were active in Washington on behalf of their constituents, but from the perspective of the cornfield or the hog lot, their main function seemed to be selling insurance, providing some other services, and maintaining a Washington office. To around 150,000 to 200,000 farmers scattered from Pennsylvania to Idaho, this was not enough. A Kansas farmer told John Bird, who was doing a story for the *Saturday Evening Post* in 1964, "I feel NFO is our last hope to keep family farmers on the land." An Iowan was equally grim. "You go to a neighborhood meeting," he said, "look around, and you wonder who's goin' to sell out next." And all the while, Bird added, the "farmers' political ranks are getting dangerously thin."

Without economic power to protect their interests, thousands of farmers were giving up. Early in 1960 the *Wall Street Journal* surveyed rural newspapers, farm organizations, and auctioneers in fifteen states and found that farm sales were mounting. In some communities the increase was 50 percent over a year earlier. An auctioneer at Webster City, Iowa, said, "I've had 50 auctions this winter compared with 33 last season." A USDA official in DeKalb, one of Illinois' richest counties, reported, "we've had close to a hundred sales . . . this winter, twice as many as a year ago." At Winona, Minnesota, an auctioneer remarked: "I've never seen so many sales for fellows who shouldn't be quitting." In one series of sales, nine out of ten of the farmers were under 40. Similar reports came from Oklahoma and Texas.

It was a matter of earning poor incomes on the farm. At Frankfort, Indiana, the secretary of the National Auctioneers Association reported that "incomes were poor around here last year and many of the younger farmers realize they'd be better off in factory jobs." Some older farmers were retiring early and taking Social Security. "Farming is getting to be a rough life," said a 35-year-old Iowa farmer who moved to town to drive an oil truck. He said that he made only $2,400 in 1958 and $1,000 in 1959. "I'm a farm boy," he said, "never lived in town until now, but my outlook here certainly is better than on the farm."

The manager of the Iowa State Employment Service in Webster City, Iowa, said that in the first three months of 1960, twenty-five farmers had been in his office looking for full-time work. One 47-year-old tenant in the Webster City community packed up his family in the spring of 1960 and headed for Oregon, where he had a job as a carpenter for $90 a week. He fired a parting shot by saying that in 1948 when he began farming he bought a tractor for $2,000, sold hogs for $20 a hundred pounds, and corn for $1.25 a bushel; twelve years later a tractor cost $3,500 while corn brought $1.00 and hogs were $12 a hundred. Another farmer said that "when prices go down you have to increase production some way," but even increased production did not help. He sold his farm and became a salesman for a farm building manufacturer. "With prices still going down this year [1960] I just decided to quit beating my head against the wall and get out." Not all farmers in all parts of the country were hit equally hard, but for increasing numbers the economic pinch was more than they could or would bear. Between 1960 and 1970 an average of about 100,000 farmers left farming each year. Even the more prosperous farmers resented the fact that their income was not growing anywhere near as fast as that of labor and business.[7]

While economic pressures were forcing thousands of farmers off the land each year, the exodus was not fast enough in the judgment of some authorities. Farm economists and other students of agriculture continued to argue that there were still too many farmers. The trouble, they said, was a surplus of manpower on the farm, and only by reducing the number of people in agriculture would it be possible to raise the income of farmers to that of the nonfarm population. In 1960 per capita farm income was only $1,174 from all sources, or 53.8 percent as much as the $2,014 earned by nonfarmers. Only by spreading the income produced by agriculture over fewer people could farmers ever expect to reach parity of income.

In late 1961 a planning group in the USDA prepared a confidential report for Secretary Freeman which recommended positive actions to speed up the flow of manpower out of farming. The proposal to reduce the number of farmers was somewhat embarrassing to the Secretary, whose rhetoric was geared to preserving the family farm. When the suggestions of this in-house study hit the front page of the *Wall Street Journal* on December 18, the Secretary had to repudiate the findings of his subordinates. The idea that such a report would be prepared in their own department angered farmers.

However, the USDA report created a mild response compared to the publication on July 16, 1962, of a study prepared by the Committee for Economic Development. Made up mainly of business executives and a few educators, the CED recommended a series of policies that would reduce the number of farmers by about one-third over a relatively short time. Current policies of price supports, acreage limitations, and subsidized exports had not worked, the committee said, because those policies encouraged unneeded resources, especially people, to remain in agriculture. What should be done was to induce a rapid movement of labor, and possibly some capital, out of farming.

The committee recommended a five-year transition period during which price supports would be phased out and as much as 20 million acres of cropland in the Great Plains and Rocky Mountain regions would be shifted to grassland. At the end of the five-year period, the free market would determine farm prices. Those who could not produce enough income at those prices would, it was assumed, abandon the farm. Other recommendations included better education for farm youth, more job information for farmers who wanted to leave agriculture, and retraining of farm people. The CED concluded that "the programs we are suggesting would result in fewer workers in agriculture, working a smaller number of farms of greater average size and receiving substantially higher income per worker."[8]

There was little that was new in the CED report. Ever since the 1930s some students of agriculture had argued that there was an excess of people in farming. In the 1950s the idea of reducing the number of farmers was often advanced as a solution to the farm problem. But because of the prominence of the CED and the committee's access to organs of public opinion, the report received more attention than the earlier suggestions. Chairman Cooley of the House Agriculture Committee quickly scheduled hearings and invited a group of witnesses who lambasted almost every aspect of the report. W. R. Poage of Texas, vice-chairman of the House Committee on Agriculture, sharply criticized the ideas advanced, and he accused businessmen, who dominated the CED, of wanting one set of economic rules for business and another for agriculture.

Witnesses representing the farmer cooperatives and the farm organizations fully agreed with the farm-oriented congressmen. James Patton of the Farmers Union testified that he was "surprised and shocked" at the CED's recommendations, and he referred to the free market for farmers as a "cult" among some people. William E. Murray of the National Rural Electric Cooperative Association called the CED study the most "coldblooded one that has ever been set before the American people." The vice-president of the NFO said that the goals of the CED were "superficial nonsense," while a representative of the Grange submitted a statement calling the recommendations "unrealistic" and a threat to farmers and to national welfare. Secretary Freeman vigorously attacked the CED proposals and pictured a declining agriculture and withering rural communities if they were adopted.

A common theme ran through most of the testimony on the CED proposals. Critics argued that to purposely reduce the number of farmers through economic pressure would be cruel to individuals and damaging to the country as a whole. As Homer Young, president of the Consumers Cooperative Association, remarked, no member of the CED was close enough to agriculture to truly know "the pain and heartache that come from being forced to abandon one's land and one's home and seek a living in strange and new surroundings." Besides, there was already considerable unemployment in the cities and opportunities for farmers to obtain nonfarm jobs were not promising.

While the uprooting of individual farmers might be a serious problem, of greater concern was the threat to rural America. Secretary

Freeman said that "a forced acceleration of this outmigration would have very serious effects on rural America." The small towns would suffer and "rural America would be irreparably changed with its communities crippled, and its institutions damaged," Freeman argued. As a banker in Springfield, Nebraska, told a reporter for *U.S. News and World Report:* "If this CED idea went into effect, it wouldn't be just 2 million farmers leaving the land. It would be the end of me and thousands and thousands of small businessmen." Jerry Voorhis, former congressman from California and executive director of the Cooperative League of the U.S.A., a spokesman for consumer cooperatives, said that the League opposed the CED program "because it would destroy the heart of our country. If that sounds sentimental," he continued, "so be it." The nation's strength, he said, "did not lie in New York, Chicago, or Detroit, but in the smaller communities of the country—communities of small property owners, many of them farmers."[9]

Defenders of rural America believed that the normal or natural attrition of farm population was bad enough. To support policies that would force people off the land at an even faster rate was considered stupid or even sinister. It was a threat to American strength and greatness. The agrarian view held that rather than urge people to abandon farming, the government ought to implement effective rural development programs that would do something to strengthen economic and social conditions in rural areas. Apart from any philosophical position, farm spokesmen especially resented what seemed to be an effort to force down further what a member of the Independent Bankers Association called the "already low price of food and fiber," and to increase the pool of unemployed farm workers to keep wages down. It appeared to farmers that the industrial and commercial interests expected agriculture to subsidize the rest of the economy—to be poorly paid toilers in the vineyards of industry, commerce, and the professions. It was not entirely coincidental that the NFO's first big strike began about six weeks after the CED report was released.

Strange as it may seem, the decreasing number of farmers seemed to have nothing to do with the continued popularity of the agrarian tradition in American thinking. The widespread acceptance of the principles of agricultural fundamentalism could be seen at every level of American society. The strength of those views provided assurance that Congress would ignore the CED recommendations—which, indeed, it did. Congress was in no mood to listen to a group of businessmen tell it how to solve farm problems. Although congressional farm policies had not been particularly successful in dealing with surpluses or low incomes, federal action had certainly placed many farmers in a better position than they would have been without the programs. In the early 1960s President Kennedy, Secretary Freeman, and the Congress seemed more determined than ever to use the federal government to prop up agricultural income.

Despite their declining position, farmers continued to maintain political influence far beyond their numbers. Although population shifts had greatly reduced the power of the farm vote by 1960, farmers still found that they could wield major influence in Washington. The apportionment of legislative districts in most states, the strength of the agrarian tradition, and the organization of various agricultural interests all contributed to this power. Indeed, farmers were much better able to protect their interests through government than by any kind of organized effort among themselves. Neither party was willing to chance alienating commercial farmers by ignoring their needs. While, as a writer for the *Wall Street Journal* observed on September 29, 1960, courting the farm vote was a "political reflex action" and a matter of "cautious protection," parties and presidential candidates in 1960 made the usual liberal promises to help farmers.

The Democrats promised to work toward "full parity income for farmers" through balanced production and consumption. In practice this meant tightening controls on output and distributing surpluses more generously at home and abroad. The Republican platform was silent on parity, but the party pledged to work for better incomes by placing more land in acreage reserves, promoting overseas sales, and distributing more surplus food through domestic programs. They did not, however, favor high fixed price supports. The main difference between the two parties, as had been true since 1953, was over the degree of production control and the level of price supports. If the Democrats expected a revolt among normally Republican midwestern farmers, they were disappointed. However, the election of John F. Kennedy was the signal to abandon the Eisenhower-Benson policy of reduced controls, and to return to greater restrictions on production and higher price supports.

The Kennedy administration promptly announced a program of "supply management." This meant getting farmers to reduce production in exchange for higher price supports. The administration also worked to increase consumption. Unless farmers cut production, Kennedy warned, the farm program would become very expensive and farm income would continue to drop. To achieve "supply management," Kennedy's advisers wanted to go beyond acreage controls and inaugurate pound and bushel quotas which a farmer would be permitted to market in exchange for higher price supports and other payments. Acreage reduction alone had never curbed surpluses because increasing productivity per acre had more than made up for fewer acres. Farmers also tended to put their poorest land in any acreage reserve.

While Congress did pass a voluntary program to cut feed grain acreage in 1961 and thereafter, lawmakers rejected the idea of quotas that would strictly control the amount that a farmer could market. The Farm Bureau led the fight against any quota plan, viewing it as an effort to place greater restrictions on farmers. Many of the more efficient farmers opposed tighter controls on feed grains because they believed they could profit from expanding their share of the market. Poultrymen and dairymen were among farmers who did not want higher grain prices because that increased their feed costs. In 1963 the administration suffered a major defeat when wheat growers turned down a plan for wheat marketing quotas. This was the first time since World War II that farmers had failed to support a quota referendum.[10]

Once farmers had rejected wheat quotas, legislation then in effect would support wheat prices at only $1.25 a bushel, or about 50 percent of parity, in 1964, compared to $2.00 a bushel in 1963. Wheat growers, however, did not believe that the federal government would permit prices to drop so low. It would have meant a loss to wheat producers of around $600 million. The farmers proved to be correct. In 1964 Congress passed legislation to support wheat and cotton prices at levels acceptable to the producers of both crops. For the portion of a farmer's wheat used in domestic consumption, the support price was set at $2.00 a bushel in return for reduced plantings of around 10 percent. Farmers also received diversion payments on that unplanted acreage. Price supports were set at 30 cents a pound on cotton for cooperating producers.

By doing something for both wheat and cotton, enough midwestern Republicans and southern Democrats united to push the legislation through Congress in 1964. However, farm representatives found that northern urban Democrats were increasingly reluctant to vote for farm subsidies without getting something concrete in return. In this case they wanted more low-cost food for poor people in the cities. In return for their support for wheat and cotton legislation, urban Democrats got votes for the first permanent food-stamp bill.

Beginning in 1939, the USDA had administered a relatively small food-stamp program, but it was abandoned in 1943 during World War II. In the 1950s, as farm surpluses piled up, there was a growing demand for some kind of a permanent food-stamp plan. Several food-stamp bills were pushed in the late 1950s, but Secretary Benson strongly opposed the idea. In 1961 President Kennedy, who had been an active supporter of food-stamp legislation when he was in the Senate, inaugurated a pilot program. Then, after overcoming much opposition from some southern Democrats and midwestern Republicans, the permanent program was enacted.

Debate over the food-stamp legislation provided a classic example of political trade-offs, the kind that farm representatives were increasingly required to make. It soon became evident that urban congressmen and senators would not support the cotton and wheat bill unless farm legislators would back an expanded food-stamp program. Congressman Cooley was quoted as saying that "we will vote for the food stamp bill, you [city congressmen] vote for the cotton-wheat bill." Republican Congressman Charles B. Hoeven of Iowa said that everyone in Congress knew what was happening. It was, he said, "the same old situation of you tickle me and I will tickle you."

Some Republican opponents of the food-stamp bill from the farm states argued that it was really a welfare measure and should be administered by the Department of Health, Education, and Welfare and not by the Department of Agriculture. They objected to the cost of food stamps showing up in the USDA appropriations when farmers received no direct benefits. Urban Democrats, such as Benjamin S. Rosenthal of New York, insisted that food-stamp administration remain with the USDA, because, as he said, that department knew more about food. Rosenthal insisted, however, that the USDA must assume its "rightful role" and be more concerned about food and not just about farmers.[11]

The major farm organizations were again divided over food-stamp legislation. The AFBF opposed it because the cost of this program would be charged to USDA appropriations, and because it would not reduce surpluses in any significant way. From the Farm Bureau viewpoint it seemed to be sufficient for the USDA to distribute surplus commodities to the states for relief purposes. On the other hand, the Farmers Union and the Grange both supported the food-stamp bill. Once food-stamp legislation was placed on the statute books, city congressmen pushed hard to expand the program. The first authorized appropriation was $375 million for three years. For fiscal 1970, appropriations reached $610 million.

As food stamps became increasingly important to urban constituents, the agriculture committees in the House and Senate which had control over food-stamp legislation found that they had a strong political weapon to deal with big-city Democrats. While some farm-state legislators initially favored HEW administration of the food-stamp program, they changed their position once they realized just how important food distribution was to urban congressmen. By keeping food stamps under control of the agriculture committees and the USDA, farm spokesmen maintained a strong bargaining position with city legislators. Not only food stamps, but the school lunch and other food distribution programs were highly important to urban congressmen. By maintaining a large degree of legislative control over the increasing flow of free and cheap food going to poor urbanites, farm leaders continued in a powerful position even as the number of their constituents dwindled.

As the House Agriculture Committee continued under the domination of conservative southern Democrats and midwest Republicans in the 1960s, the majority of the committee came into increasing conflict with liberal urban congressmen who insisted that agriculture contribute ever-larger quantities of food to welfare recipients. For years the liberals had trod gingerly in the presence of powerful agricultural leaders on the House and Senate agriculture committees, but by 1970 this was no longer the case. Urban congressmen fully recognized that political power had shifted to the cities, and they boldly demanded more free food for low-income people. Members of the agriculture committees and big-city spokesmen sparred for advantage, but ended up making the necessary concessions and compromises over programs for farmers and poor consumers.

Throughout the latter part of the 1960s, federal farm policy followed the same essential patterns that had been in effect since the 1930s. After commercial farmers rejected marketing quotas and tighter controls, Congress provided price supports and diversionary payments for those who agreed to reduce their acreage voluntarily by a prescribed amount. Price supports, however, had no relation to the costs of operation and farmers continued to suffer from a rather consistent cost-price squeeze. It was worse for some types of farmers than others, and at different times during the decade, but many producers faced strong economic pressures much of the time. Farmers may have talked about increasing their bargaining power in the marketplace, but it was still government that provided base support for many of the nation's commercial farmers.

There was increasing emphasis on food in all matters relating to farm policy. The basic laws of 1962 and 1965 were both called the Food and Agriculture Act. Whatever farmers may have thought about the situation, the expanding role of consumers in agricultural affairs had become an economic and political fact of life. Besides the domestic food programs, hundreds of millions of dollars worth of food and fiber were being given away overseas and sold for currencies which could not be converted into dollars. This amounted to $1.4 billion in 1966. The most encouraging development, however, was the increase in exports of farm commodities for dollars. Western Europe and Japan began to be especially good customers. Total commercial exports of agricultural products usually exceeded $5 billion in the late 1960s, compared to less than $4 billion in earlier years. As a result of giveaway programs, acreage restrictions, and expanding commercial exports, holdings of the Commodity Credit Corporation were down substantially by 1966 compared to a decade earlier.

Another important development in the late 1960s was the long overdue recognition that most small, low-income farmers could not be helped by the traditional agricultural programs. However much people might regret the rapid decline of farm population and the number of farms, they came to recognize the inevitable. No amount of oratory or play on the agrarian tradition could make sense out of keeping a surplus of labor in agriculture. Widespread rural poverty, which was widely publicized by Senator George McGovern, congressional committees, and the press, and dramatized on television, came to be recognized as a welfare problem, not a farm problem.

In the hearings on agricultural legislation during the 1950s and 1960s much concern was expressed over how small farmers might be assisted in becoming viable operators. By the early 1960s the emphasis had shifted to how best to help the poor, underemployed farmer get out of farming. Those who had suggested such a course of action in the 1950s had been charged with insensitivity to the needs of these people and disloyalty to the American tradition. But by the late 1960s it had become politically acceptable to propose policies that would remove farmers from rural poverty by urging them to seek other employment. Suggestions by the Committee for Economic Development that had been harshly condemned less than a decade earlier now gained surprising support.

In late 1969 Clifford M. Hardin, the new Secretary of Agriculture, stressed that the long-time attempts to adapt commodity programs to the needs of farmers on "small, inefficient units . . . had never been achieved." Hardin urged that the resources of other federal departments—Health, Education, and Welfare, Housing and Urban Development, Commerce, and Labor—focus more attention on the needs of the rural poor.[12] President Lyndon B. Johnson's war on poverty, launched in 1964 and 1965, had a rural component, but results were meager. Federal funds flowed into poor rural communities for water systems, roads, housing, food stamps, and other purposes, but little had changed for the rural poor by the end of the decade.

Reaction to the CED report a few years earlier indicated opposition to forcing people out of farming, but gradually support developed for policies that would provide off-farm employment for poor people on the land. For the most part, this meant attracting industry to low-income rural areas where people existed on the fringe of agriculture. Secretary Hardin suggested rural development and industrial employment as "an alternative for boxed-in farmers" which would still permit them to live in the country. In 1970 Senator Jack Miller of Iowa introduced a farmer adjustment bill designed to help marginal farmers make "adjustments to nonagricultural pursuits." It was "cruel," Miller said, to keep what he called submarginal producers in their current condition. While the Miller bill did not become law, it reflected a growing recognition of the realities surrounding small, poverty-stricken farmers. As industries moved into some of the nation's poorest agricultural regions, such as the Southeast, many farmers were able to abandon their effort to eke out a living in agriculture. They finally found the solution to their problems outside of farming, where it had always been for the poor, landless, undercapitalized, and nonproductive producers.

The Modern Commercial Farmer

X

During the 1960s and 1970s the nation's agriculture continued in a state of rapid change. Commercial farmers expanded their operations, substituted larger and more efficient machines for labor, embraced the scientific offerings that flowed from the agricultural colleges, experiment stations, and private industry, built new modern homes, and adopted a life-style not much different from that of city residents. The bigger family and larger-than-family farmers were actually operating businesses larger than many firms located in towns and cities. The agricultural revolution on the nation's commercial farms was at high tide.

Those who had any acquaintance with the modern farmer had long ago quit referring to him as a hick or country bumpkin. As Gilbert Burck wrote in *Fortune* as early as June 1955, "Most U.S. city dwellers know in a general way that the hayseed stereotype of the farmer is a bit out of date; just how heavily educated, mechanized, capitalized and specialized U.S. farming has become, just how swiftly the technology of the farm is moving, is something very few Americans have grasped." By the late 1960s and the 1970s, however, Americans had finally become aware of what was really happening on American farms.

On November 6, 1978, Pat Benedict, a 44-year-old Minnesota farmer, was pictured on the cover of *Time* magazine. Such national recognition placed Benedict, and by implication other large farmers, alongside many notables from government, business, and other occupations who had graced the cover of *Time* over more than half a century. "The New American Farmer," the issue's lead story, told how Benedict, who farmed 3,500 acres with some $500,000 worth of machinery, typified the modernization of agriculture and represented those who had brought such fundamental change to American farming. The *Time* writer said that Benedict was "archetypal of the farmers who make U.S. agriculture the nation's most efficient and productive industry and by far the biggest force holding down the trade deficit." More than this, Benedict was one of those approximately 500,000 farmers left in the United States who produced a major share of the nation's food and fiber. He was one of the important persons in that ever-shrinking minority.

The economic pressure placed on farmers from the 1950s onward created strong incentives to expand their operations and to increase their efficiency. They saw no other avenue by which they could raise their income and standard of living. The slogan "get bigger, get better, or get out" was accepted by more and more farmers. With no real economic bargaining power, they sought to maintain acceptable income by producing more units at a lower cost per unit. To do this, these modern farmers cultivated more acres, raised more hogs, or milked more cows. They reduced labor requirements by purchasing larger and more efficient machinery and equipment. They applied more fertilizer, herbicides, and insecticides to increase production per acre. And they specialized. These developments gave farmers the greater volume and lower unit costs which they sought. A young farmer from Milan, Illinois, put it bluntly in 1967 when he said: "You have to have the volume. To have volume you have to have large acreage. To have the acreage, you have to have the machinery."[1]

While the drive for volume and efficiency improved the income of individual farmers, it provided no solution for the overall problem of surpluses and low commodity prices. As had been true for many years, the main difficulty in agriculture was that total supply regularly exceeded aggregate demand. Nevertheless, the first farmers who adopted the newer techniques of production reaped substantial benefits. They decreased their costs per unit ahead of other farmers without increasing total output enough to depress the general price level. This group of farmers, then, improved their incomes both in absolute terms and in relation to other farmers. But as the newer techniques and practices spread throughout the entire agricultural industry, the effect was to increase the aggregate supply, which depressed prices for all producers. With falling commodity prices, the net returns of nearly all farmers tended to decline. Agricultural economist Willard W. Cochrane called this situation the "agricultural treadmill." When farmers increased their efficiency, they added to the output that depressed prices. The low prices stimulated even more efficiency, which in turn kept prices low. The term "treadmill" was appropriate in this situation, and one that farmers fully understood.

Most farmers had only a hazy notion of agricultural economics in the aggregate. However, they were all too familiar with the effect that relatively low prices had on their own operations and living standards. They looked at general economic problems from the viewpoint of their own farm and tried to devise strategies that would help them stay ahead of the game. In general, farmers had limited alternatives. As had been true earlier, many farmers or their wives took part-time or full-time jobs off the farm when they found that their farming operations did not earn enough to provide a decent living. Another approach was to struggle along as best they could, seeing their capital erode and their standard of living decline, hoping, perhaps, that they could stick it out until the Social Security checks began to appear in the mailbox. Under the most unfavorable circumstances, of course, they simply quit farming, as hundreds of thousands of farmers did in the 1950s and 1960s. However, if given a chance, the alternative chosen by most commercial farmers was to expand, to increase their efficiency, and to raise their volume of products for sale.

A farmer who wanted to expand production could intensify his operations on land already being farmed, or he could rent or buy more land. Except for enterprises such as poultry, dry-lot dairying, or confined-hog raising, there was a limit on how much a producer could increase his output on a specific amount of land. For most farmers to get bigger and increase their production, they had to buy or rent additional acres. Since several farmers in a community were often seeking to enlarge their operations, there was often a competitive scramble when land came on the market. When a quarter- or half-section came up for sale across the road or a mile or so away, a farmer reasoned that this might be his last chance to expand. Despite the criticism leveled at nonfarm investors, the greatest competition for land was usually among neighbors and not between farmers and outsiders.

Even though commodity prices did not justify the price of this increased capital outlay (that is, the land would not produce enough to pay the principal, interest, taxes, and operating costs), farmers eagerly bid for new land anyway. This may have seemed to nonfarmers like an uneconomic action. However, a farmer did not usually figure returns on a single piece of land, but on his whole operation. He could well afford to pay what seemed like an excessive price for additional acres because he spread the cost over the entire farm. For example, if an Iowa farmer bought his original 320 acres in 1940 for $100 an acre, his investment in land totaled $32,000. By 1960 he needed more land. When another 320 acres came up for sale, he bought it even though it cost $300 an acre, a figure that seemed high, but the $96,000 of added investment seemed worth it. A decade later 320 more acres came on the market, but at a price of $500 an acre. When this last land's productivity was figured in relation to current corn and hog prices, it was not a good buy. However, this purchase would give the farmer 960 acres, or enough to increase his total output very substantially. Furthermore, the larger acreage would permit him to get greater efficiency out of his equipment and machinery. When the farmer averaged the price of his 960 acres purchased over the thirty-year period between 1940 and 1970, he found that his actual investment was $333 per acre.

Calculating his outlay of $320,000 for 960 acres, the farmer found that it had been a good move, in terms of his current farming operations and in terms of his long-term investment. Besides increasing his total output, he enjoyed large capital gains as land values rose. In Iowa, farmland increased in value about 10 percent in 1965 alone, although in other years during that decade the figure was closer to 4 to 6 percent. In any event, in this illustration his capital gain on his first 320 acres amounted to about $400 an acre by 1970, for a total of $128,000, and his second 320 acres increased in value about $64,000 in a decade. With increasing land values, which gave him large capital gains, this farmer was in a strong position to buy the extra 320 acres even at what appeared to be an uneconomic price.

When farmers could not buy more land, they tried to rent or lease additional acreage. This method of expansion required less capital, but often farmers could not depend on long-term commitments from the owner. Farmers who rented land and bought more machinery and equipment to increase their operations frequently faced the loss of the land when the owner sold it or perhaps rented it to someone else. One North Dakota farmer who owned 640 acres and rented an equal amount explained his situation in the mid-1960s after losing the 640 rented acres. "There I was with 1,280 acres worth of cattle and equipment, and only 640 acres of land." Thus when 320 acres came up for sale across the road, he bought it without quibbling about the price. It was no bargain, he said, "but I was going backwards—I had to have it."[2] Another North Dakota farmer bought 960 acres in 1965 to expand his farm, even though the new acreage was eleven miles from his home place. When farmers could not purchase land nearby, they got it wherever they could within a reasonable distance. They also rented more land. Indeed, those who operated both owned and rented land were the largest and among the fastest-growing farmers in the 1970s.

In principle, many farmers did not favor the steady trend toward larger and more highly capitalized farms. "I ain't really for big farming," said a Jasper County, Iowa, farmer in 1960, but "that's what its coming to." Compulsion to get bigger came from a farmer's bank statement. An Iowa dairyman said that he would soon have to double his number of milk cows, "just to keep going." Another Iowan told a reporter that "every year you get less and less for what you grow so you've got to build your volume more and more just to hold your total income."[3]

Under these conditions, it is not surprising that the size of commercial farms grew rapidly in the 1960s and 1970s. From 404 acres in 1959, the average size of commercial farms rose to 534 acres in 1974, and continued upward in the late 1970s. There had always been a big gap between the larger commercial operators and the middle- to lower-income farmers but the differences were accelerated from the 1950s onward. In 1959 there were 336,439 farms of more than 500 acres, and they contained about 37 percent of the cropland harvested. Farmers who had cultivated 300 or 400 acres in the 1950s expanded to 600, 800, or perhaps more than 1,000 acres in the 1970s. An Iowa farmer who operated about 1,000 acres in the middle 1970s said that he could farm another section without having to buy additional machinery.

The aristocrats of these large commercial farmers were those whose cash receipts from farming exceeded $40,000 a year. In 1960 only 2.9 percent of all farmers, or 113,000, were in this category. However, this small number received 32.8 percent of the cash receipts from farming. This group more than doubled in the 1960s, totaling 235,000 by 1970. They then made up 8 percent of all farmers, but they took in a whopping 55.5 percent of the cash receipts from farming. By 1977 these larger operators numbered 510,000 or 18.9 percent of the total, and they received 78.6 percent of all cash income from farming. In other words, less than 20 percent of the farmers earned more than 75 percent of the farm income. By the late 1970s the fastest-growing category of farmers consisted of those who sold $200,000 worth of products or more annually. These farmers accounted for about 40 percent of all agricultural commodities sold.[4]

At the other end of the scale in 1960 were those 1,277,000 small commercial farmers who received $2,500 to $9,999 from farming. They made up 32 percent of all farmers and received about 22 percent of the cash receipts. Things for this group deteriorated badly over the next fifteen or twenty years. By 1977 there were about 600,000 farmers, or 22 percent, left in this category, but they only received some 4 percent of the cash receipts from farming. Most of them were part-time and earned the bulk of their income from off-farm jobs. In less than twenty years, more than half a million farms in this income group went out of business. The number of farmers who had cash receipts between $10,000 and $19,999 also came under extremely tough economic pressure. Unable to produce enough volume, thousands failed. Between 1960 and 1977 the number of farmers in this category dropped from 497,000 to 311,000. Some of these moderate-size operators moved up to larger acreages and higher incomes, but most dropped by the economic wayside. In 1978 the farmers who had $20,000 or less net annual income received over half of their income from nonfarm sources. Indeed, during the 1970s more than half of the farm families earned 50 percent or more of their income from off-farm activities.

Government programs over the years had helped to strengthen the financial position of the largest commercial farm operators. Since government land-diversion payments and price supports depended on acres and production, larger farmers who participated in the federal programs received most of the benefits. For example, USDA officials made a study of farmers who received $5,000 or more in government payments in 1967. Only 91,887 out of about 1.8 million commercial farmers received approximately $1.1 billion, or 36 percent of the total payments. The top recipient in Bolivar County, Mississippi, the Delta and Pine Land Company, got $653,252; the Campbell Farming Corporation in Bighorn County, Montana, received $166,336. The largest payment in Tulare County, California, was $257,931; in Blount County, Alabama, $93,010; in Haskell County, Kansas, $55,313; and in Coles County, Illinois, $12,585. The biggest government checks went to cotton and wheat producers rather than to corn and hog raisers.[5]

Growing criticism of large payments to farmers who least needed financial help aroused lively public and congressional debate. Defenders of the payments argued that surpluses could not be controlled unless the large and efficient producers cooperated with the acreage restriction programs. Getting their cooperation required diversion payments. Critics, on the other hand, insisted that government help to large farmers at the expense of the smaller ones actually contributed to the widening economic gap among farmers. During the 1960s political pressure built up to place a limit on government payments, and finally, in 1970, Congress set a maximum of $55,000 that could be paid to a farmer on a single crop.[6] Passage of this law reflected the eroding political influence of certain farm interests and the growing power of consumerism. At the same time, it was, as Senator George McGovern said, "consistent with the policy of encouraging family farm agriculture." In any event, this action had no noticeable effect on slowing up the growth of farm size or causing financial hardship for the big producers.

It was not easy for the American people to adjust to the reality that less than 500,000 farmers in the 1970s were producing the great majority of the nation's commercial crops, livestock, and livestock products. So far as the production of food and fiber was concerned, the United States would hardly have missed the complete disappearance of the other 1.2 million commercial farmers. By the last quarter of the twentieth century, the largest and most productive farmers had become not only a very small minority of the total population, but a fairly small and elite group within the overall farm community as well. The change had taken place with frightening speed, within a single generation.

Technology continued at the heart of the rural transformation. The equipment for planting, cultivating, and harvesting corn, soybeans, wheat, cotton, and other crops became bigger and more efficient. During the 1960s and 1970s the larger commercial farmers adopted much more powerful tractors. By the 1970s some of these four-wheel-drive monsters had as much as 200 horsepower, and more, compared to less than 40 horsepower of most tractors in 1950. They pulled eight plows and other tilling equipment through the field at up to 6 miles an hour. In the 1950s a few farmers used eight-row planters for corn and soybeans, but in the 1960s such machines became commonplace. Soon twelve-row planters came into use, and in the 1970s sixteen-row equipment was being used by some of the larger operators. These machines included electronic monitoring devices which kept farmers advised of the machinery's proper operation. With the right tractor and planter, a farmer could plant 200 acres of corn or soybeans a day. In the same operation, he distributed pesticides and herbicides. Two-row corn pickers that shucked and shelled the corn in one harvest operation were gradually replaced by four-, six-, and eight-row machines. By the late 1970s twelve-row combine corn heads had become available. These huge diesel-powered machines had automatic header controls and monitoring equipment, and moved through the field at 5 to 6 miles an hour. They could cover 80 to 100 acres a day. In corn making 150 bushels to the acre, a farmer could harvest 12,000 to 15,000 bushels in ten hours. In 1980 the John Deere Company introduced a four-row cotton picker which cost around $85,000.

The equipment for working the soil and planting wheat and other small grains also increased greatly in size. Multiple hookups for drills and discs to be pulled by powerful tractors permitted farmers to prepare the soil in as much as 56-foot widths. Much of the machinery folded up so it could be moved from field to field on roads and highways. Grain combines went from 10- and 12-foot cutting bars to 20- and even 30-foot size. Machines to cut, load, unload, and stack hay and silage made the old haying crews completely obsolete.

With advances made by engineers, plant scientists, and chemists, cotton production was completely mechanized. In 1969 a modernized cotton farmer in Louisiana described his operations in some detail. He prepared his land carefully and then planted treated seed in uniform rows with a six-row planter. He fertilized with anhydrous ammonia just before planting or in connection with the planting, depending on weather conditions. For weed control he used preemergent chemicals and later cultivated to control weeds between the rows. He ran his six-row cultivator at 9 miles an hour, covering 150 acres a day. Other chemicals were applied when the cotton plants were eight to ten inches tall, and later at lay-by time. He spread insecticides on a seven-day schedule from the time the cotton was a few inches high until October 1. Until the cotton got too tall, he used a mechanical sprayer known as a highboy, and then he turned to airplane spraying. After about 60 to 75 percent of the bolls had opened, he defoliated his cotton with aerial spray and picked it with spindle-type pickers. In the late 1960s his average yield was from 1,000 pounds of lint cotton, or two bales, to 1,150 pounds per acre, some four times what farmers were raising in the 1920s and 1930s. His production costs were $162 an acre, or about 14 to 16 cents a pound. The government support price was around 34 cents. For this producer and others like him throughout the Mississippi Delta, west Texas, Arizona, and California, cotton was a profitable crop.

In the 1960s and 1970s mechanization advanced to crops that had long resisted the use of machines. For example, a tobacco picker was available by the 1970s. Also farmers eagerly looked to machines for harvesting vegetables and fruits. A tomato picker made its debut in California in the early 1960s, and a few years later a machine was developed to harvest cucumbers. In the early 1970s blueberry growers in Michigan were harvesting their crop with a machine that vibrated the bush, caught the berries, and elevated them into a wagon. Nut producers had similar machines which shook the trees causing the nuts to fall on a large apron from which they were moved or sucked up into trucks for transport to the warehouse. In the 1970s such specialty crops as fresh peaches, strawberries, and lettuce succumbed to a degree of machine harvesting.

The reduction of labor in handling livestock was less than in crop production, but it was nevertheless very substantial. For example, large ranchers used airplanes to "ride the range," and cattle feeders adopted automatic feeding equipment which permitted the mixing and distribution of feed simply by pushing a button. Hog and poultry producers found that automatic feed and watering devices greatly reduced labor requirements. The wider use of electricity to operate automatic equipment was the key to many of these labor-saving devices.

The machines available to farmers were not only larger, but much more sophisticated and comfortable. Electronic monitoring devices, a greater number of gear speeds, hydraulic action, power steering, good lights, and other refinements increased the efficiency and ease of handling a tractor or combine. The protective cabs that became common in the 1960s provided a much greater degree of comfort. Farming had always been dirty, dusty work. A day in the field left a worker's body and clothes coated with sweat and grime. In the 1930s and 1940s, as one farmer said, a worker sat on a tractor in the open air and "chewed a lot of bugs, ate a lot of dust and got a sore rump."[7] But the modern farmer could sit in a cushioned armchair in air-conditioned comfort, and listen to the radio or to stereo music as he rode back and forth across the field.

The cost of new farm machinery rose rapidly after World War II. In the late 1940s most farmers did not spend more than around $1,000 or $1,500 for a tractor. But with larger, more powerful, and better equipped tractors on the market, prices rose steadily. In 1964, for example, a John Deere Model 3020 cost $5,500 when equipped with added features and conveniences. Prices kept going up in the late 1960s, and in the inflationary period after 1972 farm machine costs hit the stratosphere. By the middle 1970s farmers were paying $20,000, then $30,000, for a big tractor, and by 1980 as much as $60,000 to $80,000, and even $100,000, for the largest models. The same trend was true in the purchase of grain combines, cotton pickers, and other machines. It was not uncommon for the modern family farmer in the Midwest to have $200,000 to $300,000 invested in machinery; for some it was much more.

Did farmers really need such sophisticated and expensive machines and equipment? Were the increase in power, the air-conditioned cabs, the machines for harvesting fruits and vegetables (jobs that could be done by hand) really necessary? Some urban residents and a few old-fashioned farmers viewed these large and expensive machines as conspicuous consumption among farmers who seemed to be trying to outdo one another. It was true that farmers sometimes bought new tractors when the old one would have done the job. Many farmers admired their tractors and traded for new models more often than absolutely necessary. For some, a big, expensive tractor or combine was a status symbol. However, the farmer who bought an expensive piece of equipment without being able to justify it in economic terms did not stay in business long in the 1960s or 1970s.

Farmers argued that new machines were necessary to reduce labor costs. They found that labor was both expensive and unreliable, and that replacing workers with machines made economic sense. For example, vegetable and fruit growers in Michigan reported in 1970 that migrant labor costs had encouraged them to seek machines to harvest cherries, cucumbers, dry beans, blueberries, and other specialty crops that had been traditionally picked by hand. After experimenting with commercial vegetable production, the manager of First Colony Farms, a huge corporate operation in eastern North Carolina, said in 1978 that the company would only grow vegetables that could be handled with machines. Unlike most other producers, farmers could not pass higher labor costs along to the consumers of their commodities because they did not control the price. Of course, they could not pass along the increased price of machinery either, but they offset higher machine prices to some degree by getting greater performance out of the equipment. This was not possible with migrant workers. Moreover, farmers found it difficult to hire farm workers who were available when needed and who had the skills required to do many tasks and to operate expensive and complicated machinery.

The modern commercial farmer found that one main advantage of large and efficient machines was that work could be done well and quickly. Profits from crops were often closely associated with timeliness in farm operations. A midwestern corn or soybean farmer or a Mississippi Delta cotton grower might decide to replace an eight-row planter with a sixteen-row unit even though the smaller machine was in good working order. The purpose of buying the larger planter would be to shorten the planting period and take every advantage of good weather. If planting were stopped for ten days or two weeks by untimely rainfall, it could mean smaller yields and the loss of thousands of dollars. The same thing was true of harvesting. The larger combine or picker might complete the harvest before bad weather kept machines out of the field. The midwestern farmer who occasionally experienced late fall rains or early snow knew the losses that came from inability to get all of his crop in the bin quickly, or from the poor quality of moisture-filled grain. Large machinery played a positive role in reducing crop production risks.

Mechanization provided only part of a progressive farmer's continued move toward modernization. Better strains, including hybrid varieties of cotton, wheat, soybeans, corn, and other crops, helped to increase production, as did heavy fertilization and the destruction of insects with chemicals. Farming practices, too, affected output. Placing corn or soybean rows closer together and planting more seed per acre were examples of such practices. Progressive farmers developed a kind of systems approach or, as one authority said, a "package of practices" to increase productivity. They employed what the economists referred to as "aggressive growth strategies."

The modern American farmer, either directly or indirectly, continued to depend heavily on the USDA, the agricultural colleges, and the agricultural experiment stations for improved crop and livestock breeding and advice on how to get the most output per unit of input. Agricultural engineers, plant scientists, livestock and poultry specialists, and other experts constantly sought ways to increase production. Poultry scientists who specialized in nutrition, for instance, reduced the amount of feed necessary in broiler and egg output. In 1952 it took about 80 days and 3.2 pounds of feed to produce a 3.3-pound broiler; by 1977 a 4-pound broiler could be brought to market in 53 days after consuming only 1.9 pounds of feed.[8] Methods of disease control developed by poultry experts in the agricultural colleges saved producers millions of dollars annually. In the 1970s scientists at the University of Georgia's College of Agriculture and other institutions saved poultry growers millions of dollars when they conquered Marek's disease, a disease affecting poultry. As a result of improved varieties, chemical insect and weed control, and heavier fertilization, the average per-acre production of corn rose from 62.9 bushels in 1964 to 101.2 bushels in 1978. As had been true earlier, much of this increase resulted from applying knowledge developed at the agricultural colleges and experiment stations, and publicized through agricultural extension agents.

The computer age did not bypass modern, progressive farmers. In the early 1960s, as computerization moved into farming, most programs were little more than simple cost and price analyses. In the mid-1960s, for instance, some two hundred farmers were enrolled in the University of Nebraska's program for records management. A decade later farmers had much more sophisticated programs available for their use. Some of the best programs were coming out of the agricultural colleges, such as Iowa State University. Besides helping farmers view their total business operations, various agricultural and business experts set up special programs to evaluate the performance of stock and dairy cows, gains in livestock under certain feeding conditions, and the prospect of profits from a particular land purchase. In the late 1970s, specialists at Iowa State University developed some sixty programs for use with the programmable calculator that could provide quick answers to the profitability of specific crop combinations, livestock management, and other questions. An Iowa farmer who had a 680-acre grain and livestock farm said, "the real beauty of the PC [programmable calculator] is that you can keep any one factor constant (yield or price), and go up and down the scale with the other to find the areas of profitability. And it's all done in a matter of minutes."[9]

Reliance on computers indicated how important sophisticated management had become for the large, modern farmer. With investments in land, machinery, and equipment reaching $1 million, and in many cases several millions, decisions could not be left to chance. Writing down a few expense and income figures on a piece of paper when time permitted, as many farmers had done in earlier years, was no longer a sufficient data base for management decisions. As farming became increasingly capital-intensive, returns on particular operations, interest payments, cash flow, and other matters required careful attention. The farm office became a highly important room in the homes of progressive farmers. Hog farmers had to know what it cost to market a hog, corn and cotton producers needed to keep track of the relationship between fertilizer inputs and production, and every successful farmer had to know what activities made him the most money. It all added up to management. By the 1970s more and more farmers were seeing their operations show up on regular computer printouts. As one Nebraska farmer put it in 1967, "my grandfather made a good living by just plain hard work on the farm, but today hard work is not enough. You have to add management. Without it you can work your head off and still go in the hole."[10]

While much of the assistance in management and production offered to farmers came from the USDA and the agricultural colleges, these institutions continued under sharp attack in the 1960s and 1970s. Critics charged that they placed too much emphasis on improving production at a time when the main problem was surpluses. Moreover, critics insisted that most of the benefits went to the larger and more successful farmers, who needed help the least. These agencies were accused of ignoring the needs of small farmers, giving low priority to the food and nutritional needs of consumers, and forsaking the general problems of rural America. None of these charges were new, but they became more intense and attracted broader support. One of the most publicized attacks against this part of the agricultural establishment came in Jim Hightower's book *Hard Tomatoes, Hard Times* (1972). The title, "Hard Tomatoes," referred to a tough-skinned fruit developed by plant scientists which could be machine-harvested. Hightower accused state and federal agricultural agencies of ignoring human factors in agriculture, and, as Hightower put it, serving "only corporate agribusiness." He criticized the lack of help extended to small farmers and directed toward improved rural living. Deans and other spokesmen for these institutions reacted sharply, and accused Hightower of distortion, bias, and falsehood.[11]

Hightower's criticism of the land-grant college system stemmed from his basic disagreement with the recent course of American farming. He and a group that called itself the Exploratory Project for Economic Alternatives resented the transformation of American agriculture from a way of life where the emphasis was supposedly on people to farming as a business which stressed production. It was true, of course, that benefits flowing from the land-grant colleges had gone principally to the larger and more prosperous, business-oriented farmers. Their application of information and techniques coming from the experimental laboratories and fields was one of the factors helping these farmers make more money. Faculties and administrators in the land-grant colleges were trained primarily to increase production. That was considered the ultimate goal. Consequently, these individuals worked most closely with the modern, progressive farmers, as well as some elements of agribusiness, who were best able to apply their findings and suggestions.

To say that the USDA and the colleges of agriculture did not develop programs to help small, undercapitalized, and inefficient farmers did not mean that such programs could not have been devised. But even if the colleges or the USDA had developed such approaches, it would have taken a national commitment to implement them. The land-grant colleges cannot be fairly blamed for not doing what Congress and the Presidents refused to do. Most federal programs to assist poorer farmers lacked congressional support and were underfunded. For the most part, the land-grant colleges perceived their function as one of developing information and making it available to those who could best use it. They never viewed their role as that of a social agency to help the hundreds of thousands of small farmers who really needed assistance to make a living. The agricultural colleges had neither the resources, nor the know-how, nor the scope of authority to solve the basic social and economic problems of most small commercial farmers.

The modern commercial farmer in the late 1960s and 1970s had little in common with his grandfather in the 1920s, or his father in the 1940s. About the only tie between the generations was the land. The important differences between John Johnson III, now farming the Iowa land once operated by his grandfather and his father, were the acreage farmed, his volume of production, his huge investment in land, machinery, and equipment, his management responsibilities, and his income and life-style.

In 1975 modern farmer Johnson could not have kept busy more than a day or two during planting season if he had only cultivated the 160 acres of his grandfather or the 240 acres of his father. In order to use his machinery efficiently, he farmed 1,200 acres. Having inherited some of his land and bought other acreage over a period of several years, Johnson's land costs averaged out to about $500 an acre. By 1975, however, his 1,200 acres had an average value of about $1,100 an acre, or a total worth of at least $1.3 million. The figure was considerably higher by 1980, after several more years of rampaging inflation. To operate his farm, he had two large tractors, a smaller one, two trucks, a pickup, a combine with grain and corn heads, and a fleet of other machinery. Altogether, he had about $200,000 invested in machinery, plus another $30,000 for a metal building in which to house it. He had constructed storage facilities for 50,000 bushels of corn and 25,000 bushels of soybeans at a cost of another $40,000.

Besides growing corn and soybeans, he raised about 2,000 hogs a year in confinement. The hog barn and equipment, built over several years, had added another $75,000 to his total investment. Johnson's carefully kept records showed that he had invested between $300,000 and $400,000 over recent years to produce corn, soybeans, and hogs. When he added the value of his land, he found that in the 1970s he was operating at least a $1.6 million business.

Johnson raised about 600 acres of corn and 600 acres of soybeans. The cost of seed, fertilizer, herbicides, and insecticides, gasoline and diesel fuel for the trucks and tractors, his taxes, labor, and other cash outlays ran up his expenses of raising a bushel of corn to about $1.80. The expense of growing soybeans was around $4 a bushel. His annual fertilizer bill alone ran to some $35,000. While Johnson raised much of his own hog feed, he purchased a large amount of commercial and highly nutritional supplement. Finally, he paid a hired man $800 a month, plus furnishing him a house, utilities, and a couple of hogs a year for meat. He also hired some other labor at certain busy times of the year. Johnson needed from $150,000 to $200,000 a year for operating costs, to say nothing of living expenses, debt retirement, and additional capital outlays.

On the income side, his 600 acres of soybeans, which averaged 35 bushels to the acre and brought $5.09 per bushel, produced about $100,000. He fed part of his 75,000-bushel corn crop but had 35,000 bushels to sell at $2.50, which was higher than in most years during the previous decade. This gave him another $87,500. Johnson sold 2,000 hogs at an average price of $46 a hundred, or about $100 per hog, which provided $200,000, his single largest source of cash. With fairly good production and reasonable prices, Jonson's gross income reached about $390,000.

While farmer Johnson handled a large amount of money over a year's time—that is, he had a very substantial cash flow—his profit of $30,000 to $50,000 a year on an investment worth $1.6 million was not only modest but downright stingy. The 140,000 farms that sold more than $100,000 worth of commodities in 1975 had an average net income of $50,729, but the average was distorted upward considerably because it included large agribusiness operations, which in some cases sold millions of dollars worth of agricultural products. The large family farmer was much more likely to have a net profit of less than $40,000, after figuring operating expenses, interest, land, and other capital costs, and something for labor.

Whether a highly mechanized and efficient farmer such as our fictional Johnson made money from operations or lived off his capital depended to a considerable degree on how much he produced, and the price. He had no control over either of these matters. In the 1970s, mainly because of different weather conditions, average corn production in Iowa varied from a low of 80 bushels an acre to a high of 125 bushels. Variations among communities and areas were even larger. The difference of 45 bushels an acre between a big and a small crop year at $2 per bushel meant that his income varied by $95 per acre. In this situation, the trouble was that costs of production would be about the same, except for price changes in his inputs. The price of corn varied in Iowa on the average from a low of $1.04 a bushel in 1971 to a high of $2.97 in 1974 during the Arab oil crisis. When years of relatively low prices and reduced production coincided, even the good commercial farmer probably did not show a profit on operations. Many had to borrow heavily to get through the next crop year. The rapidly increasing value of land, however, gave farmers a much larger credit base. In some instances, this was the only thing that got them over critical financial crises. By 1975 farmland was selling in parts of Illinois for as much as $4,000 an acre, and more.

Farmer Johnson typified those modern farmers who seemed to be succeeding. What were the characteristics of the higher-income farmers? An economist at Kansas State University studied 324 farmers in eastern Kansas and found that those with higher incomes operated about 50 percent more land than the next income group, permitting them to spread their costs over more acres. They used more fertilizer, and had some 46 percent more gross farm income per acre. Also these high-income farmers had larger investments in machinery, which helped them keep their labor costs stable. They also paid some three times as much interest as other farmers, which reflected their high use of capital to get the volume of production they needed. These high-income farmers were also good managers.

What kind of life did modern farmer Johnson and his family enjoy? First of all, he had a modern, comfortable home. He was one of tens of thousands of farmers who had built new houses or drastically remodeled older ones in the years after 1945. It was common to see a new house on the farmstead not far from where the old home stood. Often the older house was fixed up for the hired man and his family because the farmhand no longer lived in the same house with his employer, as had been the practice before World War II. Nothing seemed to be lacking in the Johnson family's new house. Mrs. Johnson had the latest conveniences and appliances, including a microwave oven. The family room was filled with restful furniture, a television set, and stereo equipment. Farm magazines and other reading matter cluttered the coffee table. One room was fixed up for an office. Besides a desk, chair, and file cabinets, the office contained a telephone, radio, and a two-way radio for farm communications. The house was carpeted, air-conditioned, and tastefully decorated. Although Johnson usually drove his pickup truck, two cars stood in the garage for family use. The Johnsons had comforts and conveniences that their grandfathers never dreamed would come to the farm.

There was still hard work on the farm, but it was not as backbreaking or as confining as in earlier generations. Since Johnson had a hog operation, he or the hired man had to be on hand every day to care for the hogs, even though automatic waterers and feeders eliminated much of the physical labor. But he kept a sharp eye on the hogs and sold a truckload every week or two. There was always something to do around the farmstead, but the main work load came during planting and harvesting seasons. Farmers were always in a contest with the weather and they worked feverishly to get their crops in or out of the field when good weather prevailed. Those days were long, sometimes as much as 14 or 16 hours. The tractor lights often burned far into the night and looked like huge fireflies as they crisscrossed the field after dark. But the air-conditioned cab on the tractor or combine protected Johnson from the heat and dust, the seat was comfortable, and power steering eliminated the problem of aching shoulder muscles that his father had experienced after a long day of driving the old Farmall in the late 1930s.

What was a day like in the life of farmer Johnson in the 1970s? Assume that it was corn-planting time, in early May. Farmer Johnson rose about 5:00 A.M., but unlike his father or grandfather, he had no cows to milk or chickens to feed. He serviced his $30,000, four-wheel-drive tractor, and checked to make sure that his corn planter was ready to go. His wife might get up and prepare his breakfast, but Johnson often left her sleeping undisturbed, jumped into his pickup, and drove to the nearby village cafe. As he drove to town he switched on the radio to get the six o'clock grain and livestock markets. Within a few minutes he was at the restaurant and seated at a table with a half dozen other farmers, and perhaps a local businessman or two. While eating a hearty breakfast, Johnson and his friends talked about the usual things—the weather, crop prospects, and prices.

By 6:30, or a little later, Johnson was back home. After filling the planter boxes and the fertilizer and herbicide containers of his $20,000, International Harvester sixteen-row corn planter, he climbed up into the cab, pushed the starter button and moved into the field. Once under way he kept a keen eye on the electronic monitoring system, which told him if all the rows were planting properly. Moving through the field at about 7 miles an hour, he planted 20 acres an hour. About midmorning, he clicked on his two-way radio and told his hired man that he would soon be needing more seed. The seed was soon on hand and no time was lost. A little later a voice came over the radio. It was Mrs. Johnson asking him if he planned to be home for lunch or whether he would go into town for the noon meal. Since Mrs. Johnson had an early afternoon meeting, he replied that he would go to the restaurant.

It was 7:00 P.M. before Johnson made his last round through the field. He had planted 200 acres. Leaving his tractor and planter parked at the edge of the field, he drove his pickup home, stopping at the hog barn to see if everything was well there. Observing that one shoat looked sickly, he and the hired man quickly isolated the animal and gave it some medicine. Arriving at the house, Johnson took a hot shower and put on clean clothes, and by eight o'clock he sat down to supper. After supper he went to his office to return some phone calls that had come in during the day, read a letter from his son, who was away at college, and looked over some of his accounts. Later he glanced through the newspaper, watched the ten o'clock news and weather report on television, and retired to a restful night's sleep in the air-conditioned bedroom.

Mrs. Johnson's day had been much different from that experienced by her farm ancestors. After tidying up the house, putting the dirty clothes through the automatic washer and dryer, and taking a package of thick, juicy pork chops out of the freezer for the evening meal, she watched one or two morning TV shows and then worked in her flower beds. After talking to her husband on the two-way radio, and learning that he would not be home at noon, she got ready, slid into her Buick, and headed for town to buy a needed machine part and to pick up a gallon of milk, a dozen eggs, and some bread. She also met with some other members of the executive committee of the garden club to plan next month's program. Home by 5:00 P.M., she had plenty of time to prepare supper. She also had some time to work on the farm books in the office off the kitchen. On some days Mrs. Johnson spent most of her time running errands, ordering supplies, and taking care of various business matters. If need be she could drive a pickup or truck, and often did.

Only during the spring and fall of the year did Johnson work so intensely as he did that day in May. The care and sale of his hogs required either his or his hired man's attention daily, but that work could be divided in ways that gave him free time. Some of his neighbors had gone completely to cash grain farming, which made farming about a six-months-a-year activity, but Johnson believed that a successful hog operation gave him additional income security. In any event, he found time and had the money for many activities. He was a member of the board of directors of his local cooperative, and was active in the state corn-growers and pork-producers associations. He was a churchman, and his wife was a member of several organizations. The Johnsons traveled a good deal during slack work times. They often took weekend camping trips, and usually spent a couple of the winter months in Florida. They were a part of the Midwest farm rotation of corn, soybeans, and Miami! The Johnsons' life-style was no different from that of their middle-class friends—businessmen, doctors, lawyers, and others—in town.

Perhaps the greatest difference between Johnson and his father and grandfather was the time and attention he gave to management of his business. He knew the feed conversion rate and how much it cost to bring a hog to market, or the return from particular applications of fertilizer to his corn and soybean land. He kept careful records on production from various kinds of seed corn, and knew whether it was most cost-efficient to destroy weeds with herbicides or by cultivation. Johnson planned carefully. Realizing that time was money, he concentrated on the projects that would benefit him most from getting them done. He kept up with new experiments and approaches by attending occasional short courses and seminars, and subscribed to several farm business services that provided advice on management practices, farm prices, and hints on how to improve efficiency. Johnson subscribed to *Management Monthly, The Journal of the Professional Farmer* (published at Cedar Falls, Iowa), as well as several farm papers. He had become a skilled businessman in every sense of the word. Failure and bankruptcy were the grim alternatives to good management.

While modern farmers like Johnson emphasized the business aspects of agriculture, farming as a way of life was also important to them. They saw no conflict between the two ideas. Indeed, the business of farming and the degree of well-being on the land had long been closely connected. Like their ancestors, these operators had a special love for land. Working with the land through the planting and harvesting cycles gave farmers a special feeling of achievement and satisfaction. They were producing food, which was another way of saying that they were performing one of life's most important tasks. In a society whose symbols were concrete, glass, and plastic, to work with life-giving crops and livestock as an owner-operator provided a sense of deep gratification. A million or so farmers in the 1970s were able to combine a good living with the good life.

During the previous half century, however, hundreds of thousands of farmers had been unable to find a decent life on the land. They had failed to achieve that agrarian ideal of earning a comfortable living from the soil by independent effort and hard work. Even the most successful farmers could not eliminate the special difficulties and uncontrollable factors confronting agriculture. Unfavorable weather, including drought, flood, hail, wind, and frosts, was always a threat to success. As one successful young Missouri farmer said in 1980: "I can't express what it's like to stand in a dry field in the middle of July, having spent $100,000 I borrowed from the bank to get the crop in, and look down and see my soybeans burning up in the sun for lack of rain. . . . If I get three bad years in a row, I'm finished." Moreover, sharply fluctuating prices could create huge losses for even the best producers. Hog prices, for example, dropped from about $45 to $35 a hundred between 1978 and 1979, causing many producers to experience severe hardships. Livestock and plant diseases were also problems. Some large operators lost their rented acres and were wiped out when they could not get other land. Furthermore, the cost of inputs had no relation to farm prices, and farmers were in the unenviable position of not being able to pass on increased costs as did most other businessmen. Even many fairly efficient commercial farmers, as well as the less successful farmers without land, capital, or managerial abilities, had lost out because of the many factors over which they had no personal control. They quit farming voluntarily, or perhaps because they had no choice, turned to other employment, and left what they considered to be romantic notions of farm life to writers and politicians.

The modern commercial farmer was a far cry from the family farm stereotype that most politicians and social reformers talked and wrote about. Yet most of the bigger farms were family enterprises which operated entirely with family labor or with the addition of one or two full-time or part-time employees. But the growing question by the late 1960s was whether even the larger-than-family or the large family farms, of 600, 800, 1,200 or more acres, could compete with agribusiness companies and nonfarm corporations that invested heavily in agricultural production. Would American agriculture end up in the hands of a few giant producers?

If the 1960s saw a rapid growth of large family and larger-than-family farms, the increase in corporate farming more than kept pace. Between 1960 and 1966 about as many corporations began farming as a corporate entity as in the entire period before 1960. Agribusiness corporations increased their farming operations in order to acquire products for processing and marketing, while other businesses turned to agricultural enterprises in hopes of making good returns from their investment. The biggest expansion in the number of corporate farms came from the incorporation of large family and larger-than-family farms. By incorporating, farmers sometimes gained tax, inheritance, and other advantages. In any event, a 1968 survey found 13,300 farming corporations in the United States. This 1 percent of the commercial farms operated 7 percent of the farmland and had 8 percent of all farm sales in 1967.[12]

While many farmers, farm leaders, and other spokesmen for agriculture had expressed concern about corporate development in farming for many years, the problem seemed to loom larger in the 1960s. Most corporate agriculture in earlier years had been confined to fruit and vegetable production in California and Florida. During the 1950s agribusiness corporations had extended their control over much of the poultry industry. However undesirable these developments may have seemed, they did not arouse the concern stimulated by the establishment of huge corporate grain and livestock operations in the late 1960s. The Gates Rubber Company of Denver, which acquired more than 10,000 acres in northeastern Colorado, the CBK Industries of Kansas City, which bought 50,000 acres in Texas and the Midwest, Black Watch Farms, the country's largest breeder of registered Angus cattle, and Gulf and Western Company's acquisition of the Scott-Mattson Farms, consisting of some 11,000 acres in Florida, were examples of what many people considered a dangerous and highly undesirable trend.

These developments seemed to be a direct threat to the family farm, and to traditional rural communities. How could family-type farmers compete with farming operations which had the backing of industrial capital and expertise? Would not even the larger and more efficient family farms be crowded out by the corporate giants? By the 1960s these questions took on new urgency. It was not just a matter of a few corporate farms being sprinkled among a predominant number of family farms. The basic question was whether the traditional structure of American farming, where ownership and operation were lodged in the farm family, was going to be fundamentally changed to a corporate organization, where ownership, management, and labor would be separated. Could the family farm, which was oriented toward people and a way of life, meet the challenge from industrial agriculture, with its emphasis on large investments, hired labor, production efficiency, and marketing controls? More and more people were asking the basic question: Who would control American agriculture—family farmers, or huge agribusiness and industrial corporations?

The farming activities of the Gates Rubber Company in Colorado provide an ideal example of a situation that produced sharp criticism of corporate farming. In 1967 Gates quietly began to buy up land in Yuma County in northeastern Colorado for the purpose of developing large-scale, irrigated crop farming. After acquiring 10,400 acres, Big Creek Farms, the Gates farm subsidiary, purchased hundreds of thousands of dollars worth of machinery and sprinkler irrigation equipment. Managers and employees were hired. Big Creek Farms began raising corn, sugar beets, and hay in 1968, and later established a 4,000-head cattle feedlot. L. E. Dequine, Jr., manager of the Gates agricultural division, announced that the company had entered farming to diversify and to make a profit. It was not a tax gimmick, he said.

Early in 1971, only three years later, Gates officials announced that they had sold Big Creek Farms, and on February 15 the new owners auctioned off the machinery and equipment. After three crop years, the Gates stockholders refused to pour any more money into a losing enterprise. High costs, low returns, poor management, general economic conditions, and some unfavorable weather all contributed to the failure of Big Creek Farms. Opponents of corporate farming greeted Gates's withdrawal from Yuma County with undisguised glee. "I'm so damned happy Gates went under, I could holler," said one farmer. *Farm Journal* ran an article in April 1971, under the headline "Big Corporations Back Out of Farming," in which the writer summarized the failure of Gates and several other large farms established by industrial corporations.[13]

The widely publicized failures, such as Gates and Black Watch Farms, brought enthusiastic cheers from supporters of the family farm concept but did not represent a reversal in the growth of corporate agriculture. To the contrary, big farms continued to spring up in many parts of the country. In 1973, for example, Malcolm McLean, whose business interests included a large trucking firm, acquired 372,000 acres of land which extended over four counties in eastern North Carolina. This huge agricultural and land enterprise represented an investment of $60 million. The land had to be cleared and drained before it could be farmed, but by 1978 this operation, called First Colony Farms, raised 19,000 acres of soybeans and 14,000 acres of corn. It had 161 employees, most of whom were busy clearing 4,000 to 5,000 acres of additional land each year. Subsequently, First Colony Farms sold off 150,000 acres of row-crop land and uncleared timberland to the Prudential Life Insurance Company. But this was no retreat from corporate farming; it was just a shifting of land among giant businesses.

As observation confirmed the growth of agribusiness enterprises, in November 1967, Secretary of Agriculture Freeman directed the USDA's Economic Research Service to survey the number, kinds, and general characteristics of corporations involved in producing farm products. The researchers found ample evidence that agribusiness and other nonfarm corporations were expanding their investments in agriculture. Corporate farming and larger-than-family farms were producing an increasing percentage of American farm commodities. By 1974 there were some 28,000 farm and ranch corporations, which marketed 18 percent of all American farm products.

Family farmers, the Grange, the Farmers Union, some politicians, and many farm publications berated this trend which was so dangerous to the family farm. At a meeting of the Kansas Farmers Union in December 1967, one farmer jumped up and urged his 300 fellow delegates to boycott those companies which used profits they made off farmers to buy big blocks of land and then "engage in farming in competition with you and me."[14] However, neither Congress nor any presidential administration in the 1960s and 1970s was willing to place restrictions on corporation farming. Some states considered laws to forbid corporate agriculture, but except for Kansas and North Dakota there was more talk than action. There seemed to be no stopping the trend toward fewer and larger farms and the growth of corporate agriculture.

The Ever-Shrinking Minority
The 1970s

XI

ONE DAY IN NOVEMBER 1967, PAUL KLINE, A farmer from near Creston, Iowa, stopped picking corn to talk with a reporter for the *New York Times*. Standing on his corn picker, he waved his arm in an arc and said: "I can count six vacant houses from right here." The rural exodus was continuing unabated, and all of the talk about preserving the family farm did nothing to slow it down. A few weeks earlier Edward Wiedeman, who farmed near Scottsbluff, Nebraska, went over the costs of everything he purchased, and the prices he received for his products, and asked another writer for the *New York Times*, "now you tell me how in hell I'm supposed to make money!" Wiedeman said that he would "probably farm 'till I go broke." This was a variation of the old story, widely circulated in farm country, about the middle-aged farmer who was asked what he would do if he inherited a million dollars. Without the slightest hesitation, he replied: "Well, I would keep farming until it was all gone."

In 1967 farmers had experienced another one of those periodic bad years. Responding to loose talk about world famine and encouragement from the USDA, farmers planted an additional 12 million acres and brought in bumper crops of wheat, soybeans, and feed grains. The anticipated demand did not materialize, partly because of better crops overseas, and American producers were hit with lower prices. Net cash income from farming in 1967 was some $2 billion under that of a year earlier. Farmers were angry. They thought they had been "suckered" again. When Mrs. Lyndon B. Johnson visited a Wisconsin dairy farm in September she was upstaged by farmers who gathered around Secretary of Agriculture Freeman to complain over low milk prices.[1]

Despite a good deal of dissatisfaction at the grass roots, the farm question was not a major issue in the presidential campaign of 1968. Shortly before the election, Congress extended the 1965 agricultural law for one year beyond its expiration date in 1969. This provided price supports for wheat, feed grains, cotton, and other major crops through 1970, and postponed any political need to tackle the farm problem before the election. During the campaign, neither presidential candidate, Richard M. Nixon nor Hubert H. Humphrey, made a major talk on agriculture or farm policy. Both candidates, however, indicated that government action would be required to maintain acceptable farm prices. The Republicans clearly backtracked on their earlier position that the government should, as they put it, get out of farming.

Following Nixon's election, neither the administration team, headed by the new Secretary of Agriculture, Clifford Hardin, nor Congress seemed to want to take the initiative in developing new farm legislation. As usual, the general farm organizations were divided over the essentials of farm policy. The American Farm Bureau Federation continued to advocate fewer government controls, while the Farmers Union and some commodity groups insisted on production restrictions and higher price supports. By the time Congress got around to seriously considering a new agricultural bill, farmers were in another of their tightening cost-price squeezes. By June 1970, farm prices averaged only 72 percent of parity, down 3 points from a year earlier and 11 points below 1966. After much confusion and controversy, Congress passed the Agriculture Act of 1970 just before the midterm elections. The new law maintained price floors for basic farm commodities, but moved toward loosening restrictions on production. It was a compromise that really pleased very few farmers or legislators.

Urban and consumer influences were clearly evident in 1970 during consideration of the bill. Congress responded to the growing demand for some kind of restriction on government payments to a single producer when it set the $55,000 limit. Consumer power was also reflected in a more liberal food-stamp program, and in the lack of deference to such farm spokesmen as W. R. Poage, chairman of the House Agriculture Committee. When Poage suggested that people work for their food stamps, and expressed opposition to free food for "dead-beats and no-good pool hall bums," John Kramer of the National Council on Hunger responded with what one writer called "open contempt." Urban forces fully recognized that a power shift had occurred and they demanded liberal food legislation in exchange for continued subsidy and price-support programs for farmers. Less than a decade earlier, the situation had been reversed. Farm representatives were in the saddle then, and urban liberals had to appear with hat in hand to get a minimum food-stamp program. Large amounts would continue to be spent on programs for farmers, but the shift toward much greater emphasis on food policy was at hand. Within less than a decade, expenditures for food stamps alone would greatly exceed government outlays for price supports. This is not surprising in light of the fact that by 1970 probably not more than thirty to forty House members had as many as 25 percent of their constituents on the farm.[2]

As rural power declined and fewer and fewer farmers occupied the hinterlands, they complained about their growing minority status. After interviewing several hard-pressed farmers in 1967, a reporter for the *New York Times* wrote that farmers "consider themselves the neglected minority group in the country." Majority leader Carl Albert from Oklahoma wrote in 1970 that farmers had been saddled with a second-class economic position for many years because they were a too "silent minority," and W. Gordon Leith, vice-president of Farmland Industries, Inc. told a large audience in Manhattan, Kansas, in March 1977, that "the American farmer has become almost a forgotten man, the man who doesn't make much of a stir in today's noisy society."

Despite a declining farm vote, there were still points of very strong farm power in Washington. Besides the House and Senate agriculture committees, which were dominated by southerners and midwesterners, such key individuals as Jamie Whitten, Mississippi Democrat, was a towering source of farm strength. Assuming the chairmanship of the Agriculture Subcommittee on Appropriations in 1949, he had outlasted five Presidents and as many Secretaries of Agriculture. He was a vigorous defender of that ever-shrinking minority out on the farm. "Since we have gotten the news media and since we have become urbanized, few people realize that life itself is tied to the land," he declared in 1971. Whitten viewed his role as assuring farmers a "fair shake," or he said, "there won't be any people to represent."

Moreover, even though farmers were fewer in number, in a close election they could be a very influential factor in several states. They held a balance of power which they were using with increasing skill and effectiveness. By the 1970s, farmers had become one of the most active groups in American politics. This was not only a natural response by a shrinking minority to protect its interests, but can be explained by the rising socioeconomic position of many farmers. As a group, then, farmers in the 1970s were more active politically than they had been in the past, a high percentage of them voted, and they were more sophisticated in the use of political power.[3] But they still failed to work in an organized fashion.

If appropriations for agricultural programs were used as any indication, it was hard to discern the declining political influence about which farmers and their spokesmen often complained. The Nixon administration recommended appropriating about $7.7 billion for the Department of Agriculture in fiscal 1971, but Congress raised the figure to $8.1 billion. This was the largest amount ever provided in a single year up to that time for the department and its programs. Despite disunity and confusion within their ranks, the farm forces still had very formidable power in Washington.

Beginning in the middle 1960s, one of the most encouraging trends in agriculture was the rise in farm exports. The importance of overseas sales can be seen in the fact that of the approximately 300 million acres of cropland harvested, production from between 50 million and 70 million acres went into export. Unlike agricultural exports of earlier years, when much of the shipments abroad had been given away or paid for in nonconvertible currencies such as Indian rupees, dollar sales overseas were rising significantly by 1970 and 1971. Western Europe and Japan imported large quantities of wheat, feed grains, soybeans, cotton, and other products. In 1969, for example, the United States exported $5.7 billion worth of farm commodities, of which $4.7 billion was outside of various government programs; by 1971 the total had risen to $7.8 billion and $6.7 billion was for dollars.

But even with growing exports, record crops of wheat and corn in 1971 drove prices down to disastrous levels for many farmers. Wheat brought only about $1.30 a bushel, the lowest since 1943, except for 1968 and 1969, while corn plummeted to about $1 in most country markets. Hog prices experienced a big drop in 1970, and recovered only partially in 1971, leaving prices nearly as low as they had been in 1965. In light of rising prices of nonagricultural products and the intensifying cost-price squeeze on farmers, government officials concentrated on expanding exports even more. The Nixon administration talked a great deal about developing better export strategies.

Early in 1972 there were no signs that rising exports would provide a break for which farmers had been waiting. One Iowa farmer told a *Chicago Tribune* reporter that he had recently sold corn for only 93 cents a bushel. He further explained that in 1950 his corn brought $1.50 a bushel and he bought a tractor for $2,400. Now the same tractor was between $7,000 and $8,000 and corn was under a dollar. A Tifton, Georgia, farmer declared that "our politicians just don't care any more. They are more interested in cheap food for the consumer and the resulting vote from that big majority."[4] Such was the discouraged state of mind among many farmers just before a huge world demand caused exports to shoot upward temporarily.

In the summer of 1972, following a poor harvest, the Soviet Union quietly entered the American grain market and purchased some $750 million worth of wheat and feed grains. Although the Russians got the American grain at bargain prices, when the sale became generally known it had a bullish effect on worldwide demand. Prices began to rise sharply in the fall of 1972, exports shot up, and by early 1973 the American grain cupboard was almost bare. Those price-depressing surpluses had gone to Japan, Western Europe, and other markets. The USDA relaxed controls on production and farmers were not only turned loose but urged to grow more. Earl Butz, the new Secretary of Agriculture, implored farmers to plant from fencerow to fencerow. On August 14, 1973, wheat prices at Gulf ports climbed to $5.25 a bushel, more than three times what farmers had received in 1972. Soybeans topped $10 a bushel for a short time in some markets, and corn approached $3.00. Demand and prices continued strong in 1974.

For hundreds of thousands of productive commercial farmers, 1973 and 1974 were a bonanza. Many had wheat, corn, and soybeans on hand when prices began to rise. By holding their crops a few months, they got prices undreamed of when the crops had been harvested. One Iowa farmer recalled that after he sold corn in 1973, he learned for the first time what the progressive income tax was really all about. As he paid off the debt on his land in late 1973, a South Dakota farmer said, "It's just a helluva great year." The demand for staples was accompanied by rising prices for other agricultural commodities. A Georgia farmer declared that peaches jumped from $3 a bushel in 1972 to $10 in 1973, and tomatoes from 75 cents a basket to $2.50. With an unusual influx of cash, by late 1973 farmers were paying off debts, buying new machinery, trading in their old cars, and many were building new homes or modernizing their old ones. A Georgian reported: "I've fixed up my house and barn, and I'm getting new equipment. I'm living better than I've ever lived before, and I don't think this is a one-time thing." Far to the north, near Devils Lake, North Dakota, a wheat grower who raised durum wheat was receiving $8.30 a bushel for a good crop. He said he was spending $50,000 for new machinery, and had enough money left over to buy 400 more acres of land. In 1973 net farm income from farming before inventory adjustments jumped to $29.9 billion, and it was nearly as good in 1974, when it stood at $27.7 billion. Per capita disposable income from all sources for people on farms in 1973 reached $4,285, or 110 percent of that received by the nonfarm population. Up to that time, it was the only year since records had been kept that farmers had a greater per capita income than people off the farm. That was a new and exciting development.[5]

Prices high enough to make farmers decent profits contributed to rising food costs, which aroused sharp protests from urban consumers. This, of course, was nothing new. Housewives had picketed supermarkets to show their displeasure with higher meat prices in 1969, and by 1970 it was clear that consumerism had become a way of American life. Consumer power was "here to stay," said Virginia H. Knauer, Nixon's consumer affairs advisor in 1970, "with not the slightest intention of going away tomorrow."

People were probably more sensitive to higher food prices than to rises in the cost of any other consumer items. Everyone had to buy food, and they bought it often. Consequently, price increases on meat, milk, vegetables, fruit, or any other farm product were especially noticeable. Consumers also could not postpone food purchases as they could purchases of other items. As columnist Carl Rowan wrote during a later inflationary period, "inflated food prices are not like inflated prices for TV sets, washing machines, new suits or dresses. You buy those items only occasionally; you visit the grocery store every week."[6] But Americans had come to expect relatively cheap food. They spent a smaller portion of their income on food than did people in any other industrialized country—and that portion had been steadily declining over time. By 1970 Americans were spending only about 17 percent of their disposable income on food. Nevertheless, as food prices rose rapidly in 1973 and 1974 there were widespread demands that the government restrict exports of farm commodities to hold down the advancing cost of living. Urban housewives, labor leaders, and congressional spokesmen insisted that something be done to protect the American consumer.

Farmers, who had been in a constant cost-price squeeze for years, reacted angrily to complaints over high food prices and demands for export restrictions to protect domestic food prices. They were downright bitter when the Agriculture and Commerce departments, faced with hostility to higher beef, pork, and poultry prices in the summer of 1973, placed an embargo on soybeans and cottonseed. This action was soon followed by export controls on a number of other agricultural commodities. Beef was also placed under price controls in response to consumer complaints. While the soybean embargo was soon lifted, exporters were not permitted to fill more than 50 percent of their overseas orders until September, when it would be known how the new crop was turning out. In October, Secretary Butz called the embargo of the previous summer "a bad experience," and indicated that the big crop coming in would preclude any further need to restrict exports.[7]

Strong inflationary pressures in 1974, intensified by large increases for Arab oil, kept consumers continuously sensitive to higher food costs. In the fall of 1974, President Gerald R. Ford inaugurated a "voluntary cooperation system" to curb large exports of grain and soybeans in order to protect domestic supplies. While farmers were holding much of their crops for higher prices, they were bitter at this further effort to reduce their bullish export markets. They were especially hostile because Secretary Butz had assured them after the 1973 embargo that the government would not implement any more export controls. It was not until January 1975 that federal curbs on the export of wheat and soybeans were removed.

Regardless of some government controls on exports in 1973 and 1974, farmers received good prices for large crops. Their income in 1974 was down a little from the record of 1973, but most commercial farmers experienced abnormal prosperity. The old bugaboo of price-depressing surpluses was finally a thing of the past. The symbol of this overabundance since the 1930s had been the thousands of steel grain bins that held millions of bushels of wheat, corn, and other crops taken over by the Commodity Credit Corporation. In 1961 the CCC owned some 237,000 bins. By 1974 practically all of these storage facilities had been sold. This was a happy situation for farmers. They did not like large grain reserves in government hands which could be a threat over the market.

During 1973 and 1974 there were scary headlines about world starvation. Experts were talking about global famine. On August 18, 1974, the *Chicago Tribune* said that the "food situation was nearing a crisis." The previous September, widely read columnist Sylvia Porter wrote that the increased demand for food was without precedent. "All over the world," she wrote, "hundreds of millions of customers are clamoring for our production in Europe, Japan, Russia, China [and] underdeveloped lands around the globe." Porter declared that the booming demand for food marked "a watershed in world history" and was "among the most fundamental historic developments of our age." TRB wrote in the *New Republic* that "the world is going through one of the greatest economic changes in history."[8]

With such an optimistic picture of worldwide demand, and supported by the federal government, which urged farmers to tear out their fences and plant the fencerows, it is no wonder that farmers pulled out all production stops. Harvested acreage in 1974 was 38 million acres above that of 1972, and in 1975 farmers planted 44 million acres more than three years before. It had been two decades since farmers had enjoyed that ideal combination of full production and good prices. Shortages of fuel and fertilizer caused some problems, especially in 1974, but these difficulties were overshadowed by plentiful crops and profits. Cattlemen were about the only major agricultural group who found prosperity bypassing them. Most farmers believed that they had moved into a new day.

Thus it is not surprising that farmers were furious when the International Longshoreman's Association announced in July 1975 that its members would refuse to load American grain on Soviet ships until they knew what effect the grain sales to Russia would have on domestic food prices. Supporting the ILA, President George Meany of the AFL-CIO declared: "I would accept nothing less than the situation where the wheat price is not going up one penny because of this Russian sale." Here was a powerful consumer group taking direct action to protect its interest at the expense of farmers, or so it seemed to the farmers. In mid-August the Ford administration announced that it was restricting further grain sales to the Soviet Union until the effect on domestic supplies and prices could be determined. Farmers charged that the administration was caving in to the demands of organized labor. They were enraged at the thought of the government having urged them to grow more and then denying them part of their needed market. Farmers and their spokesmen considered the embargo patently unfair and discriminatory.

President William Kuhfuss of the AFBF called the boycott on loading wheat bound for Russia "nothing short of piracy," and Governor Robert F. Bennett of Kansas declared that "this arrogant action cannot and must not be tolerated by the American people." Farmers, as one Kansas observer said, were as "mad as boiled owls." "I'm damned bitter about it [the boycott]," said a Kansas wheat grower. "I'd like to know how many of them [union workers] drive foreign cars. We've got high unemployment in this country, but you don't see labor bringing its prices down."[9]

In September President Ford and George Meany of the AFL-CIO reached an agreement on the boycott. Workers would load the ships while the government renegotiated grain purchase and transport agreements with the Soviet Union. Ford told farm leaders that grain sales would be resumed when the Russians agreed to long-term grain-buying arrangements which would be designed to lessen the impact on domestic food prices. Kuhfuss complained: "we have the dilemma of a few labor leaders and the Department of Labor negotiating and speaking for farmers."[10] He called the agreement between Ford and Meany "capitulation to political blackmail." Later Kuhfuss declared, the "State Department has used farmers as political pawns in its diplomatic game through its manipulation of the market of agricultural commodities." Late in October President Ford announced a five-year agreement on grain sales to the Soviet Union. The *New York Times* editorialized on October 23 that "the agreement will prevent surging food costs from burdening consumers."

If anything was still needed to show farmers and their representatives that they no longer controlled agricultural policy, the embargoes of the Ford administration should have removed any lingering doubts. Power had shifted to consumers. The President, the State Department, labor unions, the Bureau of the Budget, poor people, and other nonfarmers were determining the major directions of farm policy. Symbolic of this change was the fact that in 1972 Secretary of Agriculture Butz had headed the American team that negotiated Russian grain purchases; in 1975 Undersecretary of State Charles Robinson led the American negotiators. The new controllers of farm policy were more concerned about broad national interests both at home and abroad than they were with the narrow interest of farmers. Indeed, they admitted to no injustice if farmers had to pay the price for maintaining relatively cheap food. Farmers and their spokesmen had to compete with a powerful "hunger lobby."

Don Paarlberg, who spent many years in the Department of Agriculture, said in 1976 that the USDA had become "more a ministry of food than a Department of Agriculture." By that time nearly two-thirds of the department's budget went for food. Thus, during the late 1960s and 1970s the initiative on farm policy gradually changed to where, as one authority said in 1976, "future food and farm policy is going to be developed outside the traditional agricultural establishment." Some of the farmers' most loyal friends in Washington clearly recognized this trend. House Agriculture Committee Chairman Poage said in 1968 that "a serious transition" had taken place in regard to the public attitude toward American farmers and their problems. It was not a partisan question, he continued, it was opposition by urban lawmakers to traditional farm programs.

The embargoes had so deeply antagonized farmers that the question of export controls became an issue in the presidential campaign of 1976. Candidate Jimmy Carter attacked President Ford for betraying farmers, and promised that, if elected, he would never approve an embargo on agricultural exports. Once safely settled in the White House, however, President Carter, like his predecessors, responded to insistent consumer demands. After experiencing a price slump for nearly three years, cattle prices began to rise sharply early in 1978. Television reporters showed housewives picking up packages of beef in supermarkets, looking at the price, and putting the beef back in the counter. Not knowing or caring about the recent hard times among cattlemen, beef eaters were up in arms over rapidly rising prices for steaks, roasts, and even hamburger. The outcry quickly reached Washington. In June the President decided to relax beef imports in hopes of slowing down rising prices. "We feel betrayed," said a Nevada rancher. Other farmers and ranchers were much less restrained in their criticism. But cattlemen had few spokesmen. The media sided with consumers. "Whatever the plight of the farmers," said *Time* magazine, "the real victims" of higher prices "are the consumers." The implication was that grocery shoppers must be protected at all costs.[11]

President Carter's appointment of Mrs. Carol Tucker Foreman as Assistant Secretary of Agriculture in charge of food and nutrition services was considered another victory for consumers and a slap in the face to farmers. Before joining the USDA, Mrs. Foreman had been executive secretary of the Consumer Federation of America. Now she would control more than half of the department's huge budget. How could producers expect much help or sympathy from such an official, many farmers reasoned. To farmers this was like putting a fox in the hen house. During the hearings on her nomination, Senator Carl Curtis of Nebraska made a big point of the fact that Mrs. Foreman had never lived on a farm and had never been connected with any aspect of farming or with an agricultural college. Curtis said that the USDA had been "created for farmers," and implied that its food distribution functions were secondary. Senator Henry Bellmon, an Oklahoma wheat farmer, said frankly that if he voted for her confirmation he would "catch 'you know what'" from his farm constituents. While Mrs. Foreman's appointment was easily approved by the Senate, it left many farmers and their spokesmen grumbling at the pervasive consumer influence in matters relating to agriculture.[12]

If farmers had taken President Carter's denunciation of farm export embargoes seriously in 1976, they learned a few years later that consistency is not considered a virtue by politicians. Because of poor crops in the Soviet Union, the Russians were scheduled to import an extra 17 million tons of grain in 1980. The prospects of such sales strengthened the sluggish market for the huge crops of wheat, feed grains, and soybeans produced on American farms in 1979. However, to retaliate against the Soviet invasion of Afghanistan in December 1979, the President announced early in January 1980 that the United States would refuse to sell most of the promised grain to the Russians. Once again farmers found themselves and their welfare embroiled in international affairs over which they had no control. Many farmers were critical of Carter's action, because, as one farmer put it, "we would like to see everyone sacrifice too." A Kansas wheat grower declared that farmers were patriotic, but "we shouldn't have to go broke being patriotic." While, in general, farmers were less critical of the 1980 embargo than many observers expected, some were, as an Iowan said, "mad as hell."[13] It seemed to more and more farmers that wherever they turned they were being asked to carry an unfair burden in such areas as consumer protection and foreign policy.

By historical standards, however, the middle 1970s were extraordinarily good for most commercial farmers. The complaints stemmed more from habit than hardship. As one farmer said, "it is just the nature of us farmers to complain." Net farm income from farming (including government payments and before inventory adjustments) dropped from a high of $29.9 billion in 1973 to $21.1 billion two years later, but farmers were still doing quite well. Nevertheless, the drop in net cash income per farm from $10,607 in 1973 to $7,617 in 1975 indicated troubles ahead.

The abnormal prosperity of 1973 and 1974 had a disruptive effect on American agriculture. Many farmers mistakenly concluded that these unusual conditions would become the norm in future years. Optimism ran high—too high. Thousands of operators bought more land, purchased large and expensive machinery, and increased their living standards. To do this, farmers, especially the younger ones, greatly increased their debts. Payments on land and machinery could be made in many cases only if prices remained uncommonly high. Thousands of producers were highly leveraged (high debt in relation to assets), and as one farmer said, they did not "leave much room for things to go wrong." A crop failure or slumping prices could spell disaster for many producers.

Unfortunately, prices began to weaken seriously in 1976, and they dropped badly in 1977. Good crops were produced both years, but demand slackened. To make matters worse, costs of production rose between 7 and 10 percent in 1977 alone. Farmers were back in that familiar cost-price squeeze. Net income was only about 18 percent of gross income, the lowest percentage on record up to that time. Operating expenses, taxes, interest, and living costs were eating up farm income. A Georgian put the situation in earthy language as he walked by his hog barn early in 1977. "This place stinks when hogs are 30-something cents [a hundred pounds]," he said. "When they're 50 cents, you don't notice it. Right now it smells pretty bad."[14]

Those with heavy debts contracted in the heady years of 1973 and 1974 were in the worst shape. They had to scramble to find money for principal and interest payments. Without reserves or sufficient equity in their land to permit refinancing, they found themselves facing the possibility of bankruptcy. A study done among four hundred bankers in nine Great Plains states in 1977 revealed that 11 percent of the farmers would not be able to pay their debts, and another 21 percent would have to refinance or sell assets to stay in business. Many of these were wheat and cattle farmers, whose main products had suffered bad price slumps. Those who still had a fairly strong capital base resented having to increase their borrowing for living and operating expenses. They believed that they deserved prices that would return a fair profit on their capital and labor. It was these boom/bust conditions of the 1970s that gave birth to the American Agriculture Movement.

During the summer of 1977 a few farmers in the wheat and cattle country of southeastern Colorado, northwest Texas, western Oklahoma, and Kansas began to talk about a farm strike as a means of forcing higher farm prices. The center of the budding American Agriculture Movement was Springfield, a small town in southeastern Colorado. The term "strike" was used because farmers believed it would be clearly understood by eastern urbanites and would attract widespread public attention. Their idea was not to plant any crops in 1978 or buy any goods from dealers and merchants unless Congress enacted legislation providing for prices at 100 percent of parity by December 14. If Congress did not act by that date, farmers said they would go on strike. During August and September the movement rapidly gained grass-roots support. At the time there was no formal organization, although interested farmers sent in small donations to an office in Springfield to pay for telephones and publicity. By early October, AAM groups were active in five states and meetings were being held in thirteen others.

Farmers gained national attention on September 22 when they confronted the new Secretary of Agriculture, Robert Bergland, a Minnesota dairy farmer, in Pueblo, Colorado. The Secretary was greeted by two thousand angry farmers, who surrounded the auditorium with tractors and pickups. License plates indicated that some of the farmers had come from Kansas, Nebraska, Minnesota, Oklahoma, and Texas. One truck carried a load of manure with a sign saying "this is agriculture's profit." Bergland told his irate audience that he did not believe Congress would be influenced by the threat of a strike, and he urged farmers to give the recently passed Food and Agriculture Act of 1977 time to help them. Farmers, however, denied that the new law would bring them any substantial relief from debt and low prices. A Texas farmer later said of the 1977 law: "They just tossed us a bone to keep us happy. We want a piece of meat."

To dramatize and publicize their demands for 100 percent of parity, farmers began a series of tractor marches. They had seen minorities such as blacks and anti-Vietnam protesters succeed with direct action and confrontation politics, and reasoned that the farmer minority might benefit from similar tactics. The term "tractorcade" was not even in the dictionary in 1977, but between September and Christmas time it became a household word. One of the earliest and largest direct-action protests was held on October 14 in Amarillo, Texas, when hundreds of tractors plastered with signs formed a procession some twenty miles long. While in the 1960s radical anti-Vietnam students had yelled "hell no, we won't go," farmers now shouted "hell no, we won't grow." Other signs were "no dough, no sow, no grow," and "we want parity, not charity." Tractorcade protests over low farm prices arose spontaneously in many farm communities.

By October the farm protest movement had spread to the Southeast, where Georgia's Tommy Kersey and Tommy Carter (no relation to the President) were among those calling on farmers to join the strike. It did not take much urging, since Georgia farmers suffered not only from low prices, but from a severe drought as well. On October 27 some four thousand farmers from eighteen counties gathered at Alma, Georgia, and drove a thousand tractors through town. The editor of the local newspaper said: "This was the biggest thing I've ever seen happen in Bacon County. These farmers are mad." An angry farmer declared, "we're fed up, we don't want to hurt anybody but we can't work for nothing." He added that Georgia farmers were very receptive to joining up with their brothers in the Midwest and the Great Plains for strike action.

Many farmers found this new approach to problem solving strange and somewhat uncomfortable. As one farmer explained at the Alma tractorcade, "I've never taken part in any kind of demonstration before. Farmers don't like to think about strikes. They're contrary to our nature." But he added that it was also against their nature "not to be able to pay our bills." The gravity of the crisis, as viewed by many farmers, clearly outweighed any reluctance to express their grievances through direct action.

Georgia farmers believed one way to attract national attention was to stage a huge tractorcade through President Carter's home town of Plains. On November 25, the nine-mile road from Americus to

Plains was filled with tractors as some two thousand farmers sought to confront Carter directly. While the President missed the excitement as he rested at Camp David, his cousin Hugh Carter was hissed and booed when he tried to explain, "Jimmy Carter understands your problem." "Why isn't he here, then," the angry crowd responded. Signs on the tractors read, "We don't want Billy's [President's brother] beer or Jimmy's peanuts," and "broke, busted and disgusted."

During the cold of December, tractorcades continued to rumble through towns and cities throughout the Southeast, the Midwest, and the Great Plains states. One observer estimated that by year's end a hundred thousand tractors had rolled in protest. Farmers said that if they did not get 100 percent of parity prices, they would not sell, plant, or buy. One of the largest tractorcades occurred on December 10, when farmers from all over the Southeast drove between five thousand and eight thousand tractors toward Atlanta. As the tractors stopped at McDonough some 30 miles south of Atlanta the evening of December 9, poised for the final trip up Interstate 75 the next morning, farmers were in an ebullient mood. One observer described it as "a combination of protest and carnival," a kind of "agricultural Woodstock." Crop dusters flew their planes over the thousands of parked tractors, dipping their wings in salute and support.[15] The next morning the tractors lumbered on to Atlanta, where they surrounded the state capitol. Some farmers drove their machines up the capital steps. A few days later several hundred tractors circled the White House and the Washington Monument in Washington, D.C. All of this activity was designed to build up support for 100 percent of parity prices and for the strike which was scheduled to begin December 14.

While the tractorcades drew national attention to farm demands, the strike date came and went without noticeable effect. In a few cases farmers tried to close markets for grain and livestock, but most of their efforts failed. Some sympathetic businessmen closed their establishments for a day or two in response to pressures from their farm customers, but otherwise it was business as usual. There were instances of minor violence, and some farmers were arrested, but the threat not to deliver farm commodities or not to buy nonfarm goods simply did not develop.

Meanwhile, several American Agriculture Movement leaders announced that farmers would carry their campaign for 100 percent of parity to Washington in January 1978. By mid-January, the strike leaders had rallied several thousand farmers in the nation's capitol. On January 18 farmers drove their tractors and trucks into Washington, causing traffic tie-ups and general inconvenience for the nation's bureaucrats. The tractors carried signs saying, "crime don't pay, neither does farming," "it's no joke we're broke," "'farmer', the most important minority in America." There were a few arrests for traffic violations. Estimates indicated that up to five thousand farmers rallied on the capital steps to hear their leaders demand 100 percent of parity. By that time AAM leaders had given up the idea of not planting crops in 1978, but urged farmers to cut back by 50 percent. Farmers also pushed their way into Secretary Bergland's office, but failed to see him. On January 20, however, Bergland met with several hundred of the strikers. He warned them not to "blow it," by destroying federal property, and urged them to work "through the political process."[16]

While Bergland had earlier said that the tractorcades were "a legitimate expression of concern" by farmers, and that government officials were "watching with sympathy and interest," neither the Secretary nor President Carter realized the depths of discontent among some farmers. If Washington politicians believed that farmers would drive their tractors around the Washington Monument, release some goats and chickens on the streets for the TV cameras, and then go home, they were badly mistaken. Hundreds of farmers were determined to lobby their senators and representatives for additional legislation.

Despite farmer complaints, however, it could not be said that the Carter administration had been insensitive to farm problems in 1977. The Food and Agriculture Act signed in September gave the Secretary of Agriculture and the President broad powers to provide price and income protection for farmers and to bring supply and demand into better balance. As farmers labored under huge crops and lower prices in 1977, price supports were raised on some crops, and the USDA rushed loan money to the neediest areas. Beginning in December, and during the first weeks of 1978, some 1.5 million government checks totaling $900 million reached farmers to help ease what Secretary Bergland called their cash-flow problems. The Farmers Home Administration and the Small Business Administration provided various kinds of emergency loans. Secretary Bergland announced that the FHA would not foreclose on any farmer, or deny him new credit, "as long as there is a reasonable chance" that the farmer could "remain on the farm." To help out some country banks that had become greatly overextended, the FHA assumed some of the shaky farm loans that threatened the solvency of those institutions.

But farmers affiliated with the AAM considered these actions completely inadequate. Still committed to a law that would give them 100 percent of parity price, they stayed on in Washington from January to April, 1978, lobbying for additional legislation. Never had farmers from the grass roots been so conspicuous and determined as when they roamed the halls of Congress, filled committee rooms, and confronted administrators. On January 24, for instance, during Secretary Bergland's testimony before the Senate Committee on Agriculture, Nutrition, and Forestry on the state of American agriculture, two hundred farmers packed the committee room, while fifteen hundred more were outside milling around in the hall.

The Senate committee also held meetings in different parts of the country where scores of farmers testified and made formal statements. Farmers expressed several major themes. Besides complaining about the hardships forced on farmers by the cost-price squeeze, they were critical of grain embargoes, the concern over consumer food prices at the expense of producers, and the general lack of appreciation for farmer difficulties in Washington. They also claimed devotion to the family farm, and warned that if food production fell under the control of giant corporations consumers would be the main losers.[17]

After several weeks of hearings, in April the Senate passed a measure that would have increased price supports and paid more money to farmers for taking land out of production. Under the influence of consumer interests and the threat of a presidential veto, however, the House refused to go along. But the farmers did not give up. After further pressure, Congress passed and the President signed an emergency farm bill in May which was considered less inflationary but still gave producers some relief. Farmers failed to get 100 percent of parity, but they did obtain some revisions in the 1977 law, more liberal land diversion payments, a kind of moratorium on foreclosures by the Farmers Home Administration, and a $4-billion emergency loan program.

Many people were surprised at the results farmers were able to achieve by exerting direct pressure on lawmakers. It may have been true, as one observer said, that farmers "scared hell out of the Senate." After all, every senator had a substantial number of farmers in his state. But the action of Congress was all the more remarkable because the regular farm organizations had given up getting any additional farm legislation in 1978. Indeed, the Farm Bureau had opposed the farm strike movement from the beginning, and President Carter did not want anything that might contribute to inflation.

In order to keep the momentum going, the American Agricultural

Movement called its first national convention in Oklahoma City for April 24. Some two thousand farm activists gathered at the State Fairgrounds to develop new ideas, to plot additional actions, and to hear Governor David Boren urge them to "keep on trying, keep on plugging. You have the support of all the people of the United States."[18] The farmers present decided to establish a permanent office in Washington and to continue lobbying for 100 percent of parity. A little later Tommy Kersey said that farmers would keep working until "we get a good farm bill passed."

Although higher farm prices in 1978—wheat in July was about $2.50 a bushel compared to $1.75 a year earlier—took some steam out of the AAM, by the fall of 1978 tractorcades appeared again. Indeed, AAM leaders declared that the 1978 tractor march on Washington would be nothing compared to the number that would reach the nation's capital early in 1979. Kersey remarked, "I imagine 20,000 tractors will pretty well shut things down."

In late January hundreds of tractors were heading toward Washington, mainly from the Midwest and the South. A Texas farmer said, "We're gonna camp out in Washington kinda like Santa Anna did at the Alamo." Most farmers arrived at the staging areas not far outside of Washington on February 2 and 3. On February 5 hundreds of tractors and pickups jammed highways and blocked the bridges leading into the District of Columbia. A line of farm vehicles some 25 miles long on one highway brought morning traffic to a halt. The police were out in force, but their cruisers were poor competition for some of the large tractors that on a few occasions rammed the police cars. Several farmers were arrested, government workers were delayed several hours on their way to work, and some high officials were late to a meeting with the President. Gerald McCathern of Hereford, Texas, the "wagonmaster" of Tractorcade II, said the activity "exceeded all our expectations."

Several thousand enthusiastic farmers gathered on the capitol steps during the afternoon. They cheered AAM leaders, who demanded that the USDA authorize producers to borrow up to 90 percent of parity on commodities in storage, and called for tariff protection and investigation of the Chicago Board of Trade. Sympathetic congressmen warned farmers to remain orderly and peaceful. When one lawmaker reminded farmers that they "didn't have any trouble last year," farmers shouted back, "yeah, and we didn't get anything last year either." Farmers continued to block traffic until police corralled most of the farm vehicles on the mall. But farmer protests continued during the week, and three tractors were burned in protest.

Neither President Carter nor Secretary Bergland would bow to farm demands. They both emphasized that farm income had risen sharply in 1978, with prices in January 1979 about 22 percent higher than they had been a year earlier. Bergland repeated that most of the farmers in trouble had made bad business decisions and paid too much for land. Protesters were incensed when the Secretary stated on a TV show February 6 that farmers simply "were driven by old-fashioned greed." In meeting with a delegation of farmers, Bergland made it clear that he had no intention of using his discretionary power to change the farm program. He also emphasized that traffic-jamming tactics had been "an unmitigated disaster from a public relations point of view," and that the farmers in Washington did not represent a majority of American producers.[19]

Farmers again took to the halls of Congress in an effort to get lawmakers to pass a resolution ordering the Secretary of Agriculture to raise commodity loan prices to 90 percent of parity. They were unable to generate support for such a move, however. With rising farm prices and even higher grocery bills, an urban-oriented Congress, as Richard Orr of the *Chicago Tribune* wrote on February 7, was "not about to pass a law that would have triggered soaring food prices."

Men and women back on the farms were more infuriated than ever at what they considered the government's cheap food policy. They agreed heartily with Congressman Bo Ginn of Georgia, who said that the "government has manipulated the whole business of agriculture to keep prices down at the grocery store." Commenting on the cost of living, Senator Edward Zorinsky of Nebraska explained that minimum wages also raised living costs, and wondered why farm prices should be considered differently.

Tractorcade II, in February 1979, had again brought dramatic attention to farm demands, but the nation's response was less supportive and more critical than it had been a year earlier. The two tractorcades indicated that farmers could still generate fairly strong support in Washington when it could be shown that commercial producers were suffering serious economic difficulties, as many of them were in 1977. But by early 1979 conditions had improved substantially for most farmers, and Congress and the Carter administration saw rampaging inflation as the country's main economic problem. It was not politically wise to do anything that would directly raise food costs. Secretary Bergland described his problem in the context of political realities. He told the Senate Agriculture Committee on January 24, 1979, just before Tractorcade II arrived, that he was trying to get farm producers and consumers to understand one another's problems. If you put farmers against everyone else, he explained, "the farmer has 40 votes in the House and everybody else has 395. Those are not very good odds [for farmers]. So I have set up strategies to accommodate the 395, to try to get them to understand the farm problems. It is a tough, hard, painful process."[20] There was general support for balancing the interests of farmers and consumers. It was an approach that appealed to basic American fairness. As the editor of the *St. Louis Post-Dispatch* wrote on February 7, 1979, in trying to "ensure fair returns for farmers, the government also has a responsibility to prevent unreasonably high grocery prices." However, many farmers believed that government action was weighted heavily on the side of consumers.

If the American Agriculture Movement failed to win widespread backing among the nation's farmers for confrontation politics, commercial farmers were generally enthusiastic about the production of alcohol for farm fuel. As diesel prices shot upward in 1978 and 1979, farmers found their cost of operations skyrocketing. Why not turn grain and other farm-produced materials into alcohol to mix with gasoline, or perhaps burn in a pure state? Farmers argued that alcohol production would use up farm surpluses and release the farmers from reliance on the Arabs for energy. When farmers began to urge alcohol production in 1978, they found little support in the Department of Energy or anywhere else. However, as gasoline prices rose to over a dollar a gallon in late 1979, and diesel approached the dollar figure, attitudes began to change. Some farmers began to build small alcohol stills, and early in 1980 farmers were considering cooperative efforts to produce their own fuel. It appeared that an idea from the 1930s would at last have its day.

The AAM, with some other farm support, pushed the idea in 1978 and 1979 that American farmers should trade wheat or corn for oil. They believed that a bushel of wheat should be worth a barrel of crude oil, the approximate exchange value before Arab oil prices began going up in 1973. "A barrel for a bushel," and "cheap crude or no food," were the new slogans. This scheme received considerable publicity, but government officials did not take it seriously. As Secretary Bergland argued, the United States did not have the same kind of control of grain exports as OPEC had over oil.

Neither the majority of commercial farmers nor the general farm organizations supported the direct action approach of the AAM. While some members and leaders of the established farm groups either openly or quietly backed the tractorcades, the old-line organizations clung to their established approaches and positions. Nevertheless, the AAM had introduced a new element into farm politics that would have to be reckoned with. Farmers saw that, despite their small numbers nationwide, they could achieve something by pinpointing pressure on Congress. The tractorcades were important mainly because they gave farmers more experience with minority and special-interest politics. Such tactics and strategies would be increasingly important in future years as a means of protecting farm interests.

While the general farm organizations and commodity groups continued to support different policy objectives in the 1970s, they did agree that the agricultural industry must find ways and means to communicate more effectively with both consumers and the federal government. In short, farmers and their spokesmen recognized that they had a difficult public relations job. This was not a new realization by any means, but the ever-growing minority position of farmers accentuated the need to get their message to nonfarmers.

Besides the stepped-up public relations activities of the older farm organizations, farmers and their representatives took additional actions to push their views. For example, in June 1973, Congressman Jerry Litton of Missouri called a "farm summit" in Washington which was attended by more than a hundred organizations that were involved in some way with agriculture. Out of that meeting, the American Council of Agriculture was formed later in the year. Governed by a board of directors which represented farmers, agribusiness, cooperatives, and others with interests in agriculture, the ACA set up offices in Washington. The staff prepared, distributed, and coordinated information which would help farm and related groups to tell their story more effectively to consumers.

Also in the early 1970s, farm women began to organize to promote agricultural welfare. On November 14, 1974, a national coalition of farm women and farm women's organizations met in Milwaukee and formed American Agri-Women. A group in Michigan known as Women for Survival of Agriculture sponsored this first meeting. It was attended by women from state "Women for Agriculture" organizations, United Farm Wives of America, located in Kansas and Missouri, and others. One farm woman declared that "bitter experience" had shown agriculture's fragmentation and the need for unity. American Agri-Women, she said, would bring an important united voice to agriculture. The organization's aim was to "magnify our political clout," and to educate consumers "to the needs of agriculture and the family farmer." To some degree, at least, farm women also wanted to counteract the growing fragmentation in the ranks of agriculture as reflected in the commodity groups. By the mid-1970s farm women's groups were active in many parts of the country, including occasional sorties to Washington. In July 1977, a group calling itself Concerned Farm Wives spent several days in Washington lobbying for higher price supports for wheat. Reporting on her stay in Washington, one farm woman said that Congress was "getting the message loud and clear."[21] By the late 1970s, farm women were making their views and influence known in an unprecedented manner.

An issue that deeply agitated many farmers in the late 1970s was the growing revelation that foreign investors were stepping up their purchase of United States farmland. While farmers had earlier been faintly aware of an occasional purchase by foreign investors, they did not consider it a serious problem. On August 7, 1977, the *Atlanta Journal and Constitution* carried a story on foreign farm operations in Georgia. There was no special reaction to that story, even though it was later shown that Georgia was among the leading states where foreign interests were buying farmland. As farmers gathered from widely scattered areas in Washington early in 1978, and exchanged information, they concluded that the purchase of farmland by foreign investors was much more common than most of them had thought. They demanded that state governments and Congress deal with the matter.

Farmers claimed that wealthy foreign buyers were forcing up the price of prime farmland, making it difficult for local citizens to purchase land for expansion. To most farmers it seemed unfair when they had to compete with a German or Dutch company, or with rich Arabs, for scarce acreage. Moreover, they argued that higher land values boosted taxes, and that foreign buyers who got special tax breaks were a threat to the family farm. Diane Brunson, a farm wife living near Statesboro, Georgia, battled hard to get some kind of restrictions on foreign purchases of farmland because, she said, "we are genuinely concerned . . . about foreigners buying up the family farms in this country."[22] She strongly urged her senator, Herman E. Talmadge, chairman of the Senate Agriculture Committee, to do something about the situation. Farmers also resented what they considered the secretiveness of foreign land purchases. Foreign investors seemed to be trying to hide their deals. When the Iowa Farmers Union sent several researchers to county court houses to ferret out foreign ownership records, they found it very difficult and sometimes impossible to discover the true owner.

In February 1978, Senator Talmadge asked the General Accounting Office to investigate the question of foreign purchases of American farmland and to report by May. The GAO gathered information from 25 counties in five states, including Georgia. While the GAO found that foreign investors owned less than one-half of 1 percent of the farmland in the counties examined, it did reveal that in one Georgia county the figure was 6.3 percent.

Senator Talmadge asked for further study of the problem, and in July 1979 the GAO issued a second report. Investigating the sales of farmland in 148 counties in ten states over the eighteen-month period ending July 30, 1978, the GAO found that foreign investors had bought 248,146 acres, or about 8 percent of the land sold in those counties. Most of the buyers were West Europeans, who concentrated their purchases in the southern and western states.[23] Even before this report became public, farmers were demanding that Congress pass legislation that would at least require foreign purchasers to reveal their holdings. Responding to farm pressures, especially from the AAM and the Farmers Union, Congress passed the Agricultural Foreign Investment Disclosure Act of 1978. This measure required foreign buyers to report the purchase of farmland at the office of the county Agricultural Stabilization and Conservation Service. A fine of 25 percent of the property's assessed valuation would be levied on those failing to file the disclosure statement within the proper time. In some states, legislatures placed restrictions on how much land a foreign buyer could purchase. In Iowa, for instance, the legislature in 1979 limited alien ownership to 320 acres.

Why did farmers and many other Americans respond so emotionally to foreign land purchases? Why was the reaction so different when foreigners invested in American farmland instead of banks, hotels, or automobile plants? The answer can be found in the special feelings that Americans had about land and farming. Many people still considered farming as something more than just a method of making a living; they looked upon it as a superior way of life which must be preserved, if not for themselves at least for others. Enduring values seemed to be associated with agriculture. If the source of those values slipped into foreign hands, countless Americans believed that the nation would be weakened. Foreigners must not, they said, be permitted to acquire this precious resource which provided the roots of the American character. There were a sizable number of farmers, of course, who did not let their emotions stand in the way of profit. As one Georgian said: "Let's don't let them [foreign investors] get by with buying our land, but don't stop it yet, not until I can sell mine."

A more fundamental question that foreign ownership of American farmland raised in the late 1970s was the future structure of American agriculture.[24] The growing number of large producers, the expansion of agribusiness, and the decrease in the number of traditional family farms all raised the issue of the kind of farming that would be, as Secretary Bergland put it, "in the ultimate best interests of farmers and the nation." This, of course, was not a new problem. It had been discussed by farmers, economists, politicians, and social critics since the 1930s. The Food and Agriculture Act of 1977 had directed the Secretary of Agriculture to investigate the structure of American farming and to report back to Congress. It was not until 1979, however, that any public action was taken. On March 12, Secretary Bergland brought the problem of farm structure to national attention in an address before the National Farmers Union convention in Kansas City.

Bergland declared that it was time to consider where American agriculture was and where it seemed to be going. He expressed concern over current trends in farming and the resulting "distortion" in "the traditional rural social order." The Secretary felt that something of "lasting worth" had been lost, not only in the countryside, but in "urban America as well." Bergland said that he did not want to see a "handful of giant operators own, manage and control the entire food production system." Yet, he added, that was where "we are headed, if we don't act now." He asked farmers and their leaders to begin a national dialogue on the future structure of American agriculture.

Following Bergland's challenge, meetings were held throughout the country in late 1979 to seek the views of farmers and their spokesmen. The plan was to build a base of ideas, information, and recommendations on which policy makers could rely when Congress dealt with farm legislation in 1981. In January 1981, the Department of Agriculture published *A Time to Choose: Summary Report on the Structure of Agriculture*. There was nothing new in the report's conclusions that changes in agricultural marketing, improved technology, and government tax, credit, and commodity policies had stimulated the rapid surge toward larger and larger farms. Nor did the recommendations break any new ground. These called for government action that would facilitate intergeneration land transfers, modify tax policies affecting farmers, provide changes in commodity programs favoring smaller producers, and direct research toward the special problems of "the small and medium-sized farms."

The main trouble with the dialogue on farm structure was that it came nearly a half century too late. The framework of American agriculture was firmly established. It was not likely that it could be changed. If different tax, credit, and price-support policies had been employed a generation earlier, the development which resulted in the 200,000 largest farms producing nearly two-thirds of all agricultural output might at least have been retarded. By 1980, however, neither the national will nor practical political power existed to change the situation. The national dialogue encouraged by Secretary Bergland provided an outlet for emotional expressions about the family farm and the importance of the nation's rural heritage, but there was little chance that it would significantly modify the structure of America's agriculture. The time in history when a different structure could be achieved under the democratic process and private enterprise system had, for better or for worse, passed.

Views from the Grass Roots

XII

How did farmers view their growing minority status and the changing structure of American agriculture? Between 1976 and 1980 about 250 commercial farmers scattered from Virginia to Washington State and from Georgia to Montana were asked to answer a questionnaire that solicited their opinions on a wide range of current issues. Sixty percent of them responded. In 1978 approximately 340 farmers in the Midwest answered a poll distributed by the *Farm Journal* which provided added insights into farm thinking. Other farmers were interviewed in varying and widely scattered sections of the country. While the results do not represent a true cross section of farmers or a scientific poll, the responses represent a large body of informed farm opinion.

In reply to the question as to whether farmers thought of themselves as a distinct minority, the answer was an overwhelming yes. Some 85 percent of the farmers polled by the *Farm Journal* considered themselves a member of a minority group, while the broader national sample showed that 70 percent recognized their minority status and its consequences. Although viewing themselves as a minority, they believed, as one farmer wrote, that they were a very "respected minority." An Illinois operator declared that farmers were "one of the most important minority groups in the world," while another saw farmers as "an elite minority group."

Not all farmers, however, saw themselves as part of a minority. Although farmers knew that at the time they made up only about 3 percent of the total population, the environment in which they lived made it difficult for them to realize their minority status. The American landscape was still predominantly rural, and farmers worked in a society where they and farm-related businesses and services were extremely important. So while farmers gave intellectual assent to their minority situation, they lived, worked, and socialized mainly with other farmers, making it hard for them to grasp their true status. A Kansas farmer wrote that he did not think of himself as a member of a minority group, because "most of the people I know are farmers," while a Georgian rejected the minority label by saying "middle Georgia [where he lived] is a farm section." What farmers knew and what they personally experienced produced contradictions in their thinking, and added to the problems of achieving any unity in their thought and action.

Having fallen into a minority status, between 73 and 79 percent of the farmers polled believed that they suffered discrimination at the hands of other and more powerful groups. By an overwhelming majority, farmers said they were not properly represented in the United States Congress, that federal farm policies had not been fair and helpful to them, and that consumers had gained far too much influence in agricultural matters.

Many farmers sharply criticized the federal government. In answer to the question posed in 1978, "Do you believe that the farm policies of the federal government have been fair and helpful to you as a farmer over the last few years?" 80 percent of 319 midwestern farmers replied no. In a broader national sample of 150 farmers, 65 percent gave the same reply. An Idaho orchardist was critical of federal restrictions on the importation of Mexican labor, and wrote that "our white people don't want to pick fruit" because they received food stamps and welfare checks. A Kansas farm wife said there were far too many federal restrictions and regulations. An Illinois farmer wrote that farm production had become too important in the eyes of the bureaucrats to let the agricultural industry manage itself. He then cited the restrictions and costs of the Occupational Safety and Health Administration and the Environmental Protection Agency. One Iowan who had been farming for fifty years declared, "we don't need OSHA or someone who has never seen a farm to tell us how to operate it."

Farmers were bitter over government policies which they believed contributed to their inflated operating costs in the late 1970s. One farmer wrote that the most important problem for farmers as a group was the "inflated prices of machinery, parts, fertilizer and fuel." A successful South Dakota farmer-rancher said that government loan programs "have increased and inflated the price of land as many young farmers buy land because they can get a loan, and the only way it will pay is by more inflation." Another farmer said that government policies supported the poorest operators and created competition for the most efficient farmers. "The government," he wrote, "tends to take from [the] successful and give to failures." Labor unions were blamed for much of the inflation and rising costs of operation. As one farmer put it, if the government could have controlled labor unions "20 years or so ago—things could have been better." A Georgia dairyman complained that federal actions were not fair to farmers because "too many farm policies are made by people who have no idea of what it takes to operate a farm today [1977]."

Farmers were most critical of the federal government in the areas of export restrictions, the importation of competing agricultural commodities such as meat, and the changing and uncertain price policies. "The embargo and dock strike was [*sic*] very unfair," said one farmer, while another wrote that "there was too much government interference in pricing." Complaining of federal action, one operator accused the government of "trying to drive our prices down."

Farmers were very unhappy at their representation in Washington. The fact that for the first time in history both the President and the Secretary of Agriculture were genuine dirt farmers did not change this situation. Indeed, the fact that President Carter, a successful Georgia peanut farmer, resided in the White House, and Secretary Bergland, a Minnesota dairyman, headed the USDA, caused farmers to be more critical than usual. They expected Carter and Bergland to treat their problems more sympathetically and to respond to their needs with what they considered greater understanding. But farmers claimed that neither the President nor the Secretary provided useful leadership. A Kansan observed that the USDA was not of much help to farmers because it had become "more of a consumer watchdog instead of a farmer bulldog." Moreover, according to many farmers, government could not be trusted. They cited the instances when federal officials urged farmers to increase production to meet world food demands. After producers responded and surpluses again developed, the government failed to maintain profitable prices. That, farmers said, was letting them down.

It was in Congress, however, where farmers saw their influence slipping away. Among one group polled, 89 percent said their interests were not properly represented. Many farmers expressed the idea that politically they just did not count for much anymore. A large Kansas operator wrote that "the farm bloc is dead—farmers don't possess enough votes to be considered a political force. So agriculture is subject to urban political views." An Idaho farmer said: "simply put, the congressmen and other politicians go where the votes are," and that was not to the farm. A Texan bemoaned the fact that "farmers do not have enough votes to influence the politicians," while a Middle Georgian observed that "some of us think we are being sold out." A Nebraskan put it bluntly when he declared, "there are too many urban legislators." One midwesterner grumbled that "politicians go where the votes are. A man capable of managing a million-dollar business counts no more than a dribbling idiot at the polling place."

Farmers repeated the theme of poor and inadequate representation time and again. A South Dakotan who had farmed for thirty-six years said that "at election time our farm vote doesn't mean a thing." Another South Dakota farmer objected that he was represented by what he called political deadbeats who "didn't know what end their head is on." A Minnesotan said that congressmen paid no attention to farmers "unless we do a lot of bitching and people get tired hearing that." From a large Georgia operator came the observation that "even most of the senators who have farm backgrounds have lost all touch with today's agriculture," while a neighboring Alabama peanut and soybean farmer wrote that "a great majority of those elected do not understand agriculture." An Indiana farmer claimed that Congress contained "too damn many lawyers." The embargoes, said an Illinois operator, provided added proof that farmers did not have strong representation in Washington. A Montana rancher blamed the "one man one vote" principle for policies which hurt the agricultural producer.

Where had the source of political power shifted? In the view of most farmers it had moved to the ranks of organized workers. Labor unions were seen as the pampered darlings of politicians. An Illinois operator said that politicians were "elected about 70 percent by labor votes so they cannot adequately represent us and get elected again." A large South Dakota farmer declared that "the cheap food policy is designed to help the laboring class in return for their votes." Another producer said that workers wanted high wages "to make our equipment and cheap food so they can buy boats, snowmobiles, motorcycles, campers etc." Many farmers were extremely hostile toward George Meany and the longshoremen at the time of the 1975 grain embargo. "I get so damn disgusted over the array of strikes for any purpose," wrote one farmer, "usually accompanied with no increase in productivity, that I am ready to endorse firm action." Farmers saw what they considered high union wages adding to the cost of machinery and other production needs, and union votes as the source of strength for the government's cheap food policy.

While believing that nationwide agriculture had poor spokesmen and ineffective representation, a majority of farmers were satisfied with the performance of their own congressmen and senators. This seeming contradiction has been confirmed by other surveys. Some 57 percent of 343 farmers in the Midwest reported in 1978 that they wrote to their congressman or senator once or twice a year when they were unhappy with government actions, and 55 percent got what they considered satisfactory responses. However, beyond occasional letter writing, most farmers in this group were not active politically. Only 28 percent said they were active in campaigning, canvassing for votes, or other political activity.

If farmers were pessimistic over the decline of their political power, did they see any hope of maintaining strength and influence through their general farm organizations or the farmer cooperatives? The answers from 78 to 80 percent of those surveyed who were members of a farm organization were mixed, but comments by farmers indicated that agricultural organizations were not making a major difference in farmers' economic welfare. Of the 72 percent who reported that their organizations had been helpful, most viewed their organization's political activity as the most important contribution. Farm organizations played "a key role in the political arena," wrote one farmer. "Individual voices are seldom heard in government as it works today." As an Illinois farmer wrote, the Farm Bureau, his organization, was his "voice in Washington." Another Illinois operator said that the main value of the farm organizations was "lobbying for farmers," while an Indiana farmer observed that farm organization representatives watch the "bureaucrats and federal register to keep us informed—we don't have time to read all the junk." A South Dakotan reported that his organization was important in "keeping me informed of what the government is trying to do to me." Another producer said that the organizations helped to "stave off injurious legislation" and helped "to pass beneficial legislation." It was the educational and informational functions of the farm organizations that farmers believed were most valuable to them.

But farmer after farmer explained that their organization provided very little real help. A member of the Kansas Farm Bureau wrote that the organization issued statements and proclamations, but seldom accomplished much. He maintained his membership, he said, "for vehicle and farm insurance only." Another Kansan said that the Farm Bureau "is an insurance operation here." A Minnesota member of the Farm Bureau declared that his organization was more concerned with selling insurance than in organizing to deal with basic economic problems. A South Dakota farmer-rancher thought that the farm organizations tried to be helpful but concluded that their influence was minimal. Many farmers who were members of the Farm Bureau and the Farmers Union saw their organizations as little more than insurance companies. In a few cases Farmers Union members argued that their organization fought effectively for such things as parity prices, preservation of the family farm, farm price legislation, and other measures of positive value to both members and nonmembers, but this was not the general view.

The inability of farm organizations to unite and speak with one voice for agriculture concerned many farmers. They realized the desirability and need for such unity at a time when farmers were becoming so few in number, but no one predicted that the main general farm organizations would ever get together and provide one spokesman for agriculture. Farmers saw clearly why this would not happen. They recognized that the interests of farmers were too varied and sometimes contradictory to permit unity of action.

Farmers had even less confidence in their cooperatives being able to provide much leverage in the marketplace, or to help them improve their economic position. Seventy-six percent of the farmers in one survey and 78 percent in another reported belonging to a producer or consumer cooperative. Some of these farmers were very enthusiastic about cooperatives and how they could help farmers. Their main value, as one Minnesota farmer wrote, was that "they keep a lot of other businesses in line." In other words, cooperatives provided another element of competition in markets that were increasingly influenced or controlled by monopolies. Farmers were saying that cooperatives helped to keep conditions from getting any worse than they were on the farm. But the most remarkable thing about farmer replies was the number belonging to cooperatives who frankly admitted that they received no benefit from their associations. One co-op member wrote rather cynically that his association gave him lower prices for his crops and charged him more for his supplies than noncooperative business. This attitude may be explained by the fact that the surveys covered farmers who were generally among the most successful. Charles Shuman, president of the American Farm Bureau Federation from 1954 to 1970 and well-to-do Illinois farmer, judged that the big and most prosperous farmers probably benefitted the least from membership in cooperatives and the general farm organizations. In any event, there were relatively few farmers who saw cooperatives as a way for farmers to increase their bargaining power.

The thing that disturbed and irritated farmers more than anything else was what they considered the predominant power of consumers. Ninety-five percent of those surveyed believed that agricultural policy was becoming too heavily influenced by consumer interests at the expense of farmers. This perception, combined with the idea that consumers did not care about the welfare of farmers and were only interested in cheap food, frustrated and angered men and women at the grass roots. A Texas farmer was only one of many who referred to the appointment of Carol Foreman as Assistant Secretary of Agriculture as convincing evidence that consumers got first priority in government farm policy. "Consumers," wrote another farmer, "are insensitive to farmers' problems because they want cheap food and believe it is their right to cheap food from the policies prepared by Carol Foreman and USDA." Farmers were weak, said a Colorado producer, because they lacked economic and political power. Labor would make a decent profit or strike, he continued, but consumers do not believe farmers can strike and "neither does the government so they [*sic*] force food prices down." An Alabama peanut and soybean producer said that the most important problem facing farmers was the "uneducated consumer influence dictating farm policy and pricing." The Senate and the House, wrote a South Dakota farmer, "have adopted a cheap food policy," while an Idaho farmer and official in the state department of agriculture said that the USDA "seems more interested in improving their [*sic*] image for consumers rather than family farmers."

Farmers, of course, saw the cheap food policy operating at their expense. A midwesterner remarked that "most consumers I have talked to don't care about our problems and think we should raise their food at low cost." A South Dakotan wrote that "everything is geared to keep the consumer happy. The consumers want cheap food and they don't care at whose expense." "The New York housewife" was the only person listened to, according to a South Dakota critic. Some farmers were even more blunt. When asked whether consumer influence was too strong in determining agricultural policy, an Ohio farmer replied "hell yes," while an Idaho seedman answered "sure as hell."

Some farmers, of course, took a more dispassionate view of the situation. Although consumers did not like high prices, one farmer said that he believed nonfarm residents were becoming more informed about agriculture and the need for farmers to make a profit. Others declared that farmers, as well as town and urban people, were also consumers and that all groups had mutual interests. But 64.5 percent of midwestern farmers in the 1978 *Farm Journal* survey believed that farmers and consumers were natural competitors with sharply different interests. Indeed, farmers viewed urban consumers as the main force at work beating down farm prices.

The most persistent complaint from farmers was about low prices, meager or nonexistent profits, and the almost constant cost-price squeeze. One farmer wrote in 1977: "I worked my butt off this year and didn't break even. Not because I didn't raise anything but because I can't get a fair price." In response to a question on whether the most successful farmers in the community in the late 1970s were making 6 to 9 percent profit on their investment in land, machinery, and equipment, 91 out of 142 widely scattered farmers replied no. A Kansas farmer said in 1977 that he believed only about half the farmers he knew were making that level of profit. The economic pressures, he indicated, were especially strong on small operators. "I really think the small 'family farm' is already done for—larger operators only will continue to make it," he explained. An Idaho farmer wrote that "farming has never paid for the commensurate skills and risks involved." An Illinois producer reported that he made enough to live on comfortably, but that he was not receiving a decent return on his investment. Farmer after farmer commented that they needed better profits—or just some profit—so they did not have to live off of their land equity.

Many farmers reported that increasing land values was all that kept them afloat. They were fully aware that under good bookkeeping practices operational profits were closely associated with *when* a farmer had acquired his land. An Illinois resident reported that farmers in his community were making up to 9 percent profit, but only if they bought their land at preinflation prices. A Kansas farmer explained that the answer to the profit question depended "on when the land was purchased." Complaining of low returns, a South Dakotan wrote: "if I sold out, my interest on my investment would be a lot more than I'm making." But, he added philosophically, "I like farming and will stay with it." A Kansan who cultivated over a thousand acres said he was making less than 6 percent profit on his investment, but added: "I still enjoy what I'm doing but for how long I don't know." Replies confirmed that during the late 1970s it was more profitable to own land than to farm it. That is, the return from ownership exceeded that from operations.

The farmers who had what some called an "equity head start" were clearly in the best financial shape. Farmers who acquired land in the 1940s and 1950s realized a huge capital gain by the late 1970s. With phenomenal increases in land prices, the farmer who acquired only a few hundred acres in some communities found himself moving into the millionaire class or more between the 1950s and 1970s. This situation was not, however, without its problems for farmers. Higher land values brought added current real estate taxes and increased estate taxes later.

If farmers complained about poor profits and low incomes, did they believe they could improve their situation by getting together and gaining control over their prices? Some 65 percent of those answering this question said they saw no prospect for such a development. A few did not even consider organized marketing power desirable. Answers to this question reflected a very deep strain of farmer independence, and farmers saw such action as a move toward giving up their freedom. A small South Dakota operator said that the farmer was the most "self-dependent person alive. Somebody tells him to do something and he is just going to do the opposite." A farmer, this operator continued, is his own boss and "that is one of the most important reasons he is a farmer." "I will give up gold for freedom," said an Illinois farmer. "As long as we are farmers," explained an Iowa farm wife, "we will not get together." A Georgian admitted that the best answer to farm problems might be collective buying and selling, but "deep down I believe in every man being able to make his own decisions." It was that kind of independence, he added, that made men out on the land say, "I'm proud to be a farmer." An Illinois operator said that if farmers ever got together to set prices, "they will not be American farmers." The implication was that such action would verge on being un-American.

Others viewed the situation somewhat differently. A South Dakotan declared that farmers were "too greedy and independent to organize." An Idaho agricultural official and farm operator observed that farmers would not get together because "they're too damn independent and would rather starve separately, rather than prosper collectively." A Kansan remarked that you "can't find two farmers who will agree on everything." A Nebraska farmer who had been in business thirty-five years wrote that "most of us farmers live on farms because we like the way of life, . . . be your own boss. And therefore we'll settle for less income than the factory worker, the professor, government employee, but there is a limit [to the sacrifice]."

But farmers saw the practical difficulties of achieving any effective organization that might give them power to bargain over agricultural prices. Many operators mentioned conflicting interests and competition among producers as major factors that kept farmers apart. Besides being too diverse, said an Alabama farmer, "they are too spread out geographically." He also mentioned what economists had pointed out for years, that "farmers could not economically do some of the things required to control prices," such as "withhold products from market." A Colorado farmer-rancher said that he did not believe farmers would get together, "because each farmer is a producer by himself and what helps others may take away from him." Farmer cooperation to set prices could be achieved, commented a large South Dakota operator, but "it won't happen. I feel we will keep working harder and longer until we have to sell out." Some who thought farmers should unite said it would not happen because farmers were, as one person put it, "too damn dumb," while another exclaimed "there are too many stubborn and ignorant ones."

Most farmers agreed that only after the number of producers was reduced far below the figure of the late 1970s would they be able to combine and set prices; only after the number of farming units was reduced much further, they believed, would they be able to act in unison. A South Dakota rancher explained that when large corporations gained control of American agriculture, producers would govern their markets. Only after the United States developed "more of a corporate agricultural economy" would farmers ever get together. However, when corporate ownership and agribusiness controlled production and markets, it was recognized that very few remaining operators would fit the definition of the traditional family farmer. When that day came, an Oklahoma rancher predicted, "prices of foodstuffs will be based on cost of production plus a fair profit." Many farmers insisted that consumers had the most to lose if agriculture ever fell under the control of big business. Corporate pricing policies, one Iowan said, would cause food prices to skyrocket.

The independence of farmers was contributing to their own economic destruction. Most of them fully recognized that they operated in a highly organized economy, which placed them at a sharp disadvantage, yet they continued to resist any kind of union or cooperative action that would effectively increase their economic power. The rest of the economy did not need to develop strategies to weaken or reduce farm influence. Farmers themselves did that. "If farmers can't swim," said one strong individualist, "let them sink." Many were sinking in the 1970s.

If farmers indicated little support for uniting among themselves, what were the prospects for forming a political coalition with one or more other groups in American society? That was one obvious way to increase farmer power. Moreover, this idea had a long and honorable, although unsuccessful, history in the United States. When midwesterners were polled on this question, they divided almost evenly for and against any kind of coalition with nonfarmers. Of the 50 percent who did favor the proposal, 44 percent believed that farmers could best team up with labor, 60 percent preferred trying to work with business, and others favored both labor and business.

The farmers who wanted to cooperate more closely with labor expressed admiration for the power unions had achieved. While many farmers were anti-labor, those who favored some type of strong organization believed that labor had established the correct pattern. One midwestern producer said that farmers should make a coalition with the Teamsters Union. That, he said, would produce results. Favoring cooperation with workers, another farmer wrote that "unions have strong power and have helped the laboring man." But many farmers had deep qualms about trying to work with labor unions. While admitting that cooperating with labor might bring the best rewards, one farmer admitted that labor was the group "I would least like to work with." For the most part, farmers believed that workers had interests contrary to their own, and many farmers were solidly anti-labor. The welfare of farmers and workers appeared contradictory. As consumers, labor wanted cheap food, while farmers demanded higher commodity prices. One farmer explained that "high agricultural prices and low food prices don't go together." "Teaming with labor," said another operator, "would be our worst mistake." One anti-labor farmer put it more descriptively when he said, "union labor are [*sic*] like fleas on a pup."

Rather than trying to find allies in labor unions, the majority of farmers who favored any kind of coalition believed that it should be arranged with business. The larger and more successful farmers, especially, held that they had more in common with business than with labor. Their more natural allies, they said, were the machinery, fertilizer, petroleum, feed, and other companies with whom they dealt. "Each farmer is a business person and has the same problems," wrote one farmer, while another said, "farming is big business." A coalition with business seemed more feasible to some farmers, because, as one said, business stood to lose "as much as the farmer does if he can't make it." Business and agriculture, explained another farmer, "have a common interest in preserving an economic environment in which free enterprise can flourish." A farmer "is a businessman," wrote another operator, and added that "if he isn't he will soon be out of farming." It seems clear that as farms became larger, more highly capitalized, and produced a higher gross income, the operators identified more closely with business than with other groups.

Farmers were deeply concerned about many other issues in the late 1970s. Many of them believed that what farmers needed was more effective public relations. As a Montana rancher remarked, farmers would "get a better shake if the farm story were more fully told and fairly interpreted by the press." A Michigan farm wife who was active in the Agri-Women movement declared, "we have to speak out and . . . educate to create understanding of the uniqueness of agriculture and its complexity." Others emphasized the importance of agricultural exports and the significant part they played in helping to pay for imported oil. Some farmers wanted to stress the miracles of American farm production and publicize that one American farmer in 1978 produced enough food to supply 65 other persons. Farmers boasted of their contributions to the national welfare, and had a very positive self-image. They were equally certain, however, that they were not properly appreciated by the rest of American society. Even discounting the prevalent "poor farmer" syndrome evident throughout much of American history, many intelligent and informed farmers believed that most other citizens did not understand the peculiar and complex problems that faced agriculture.

Any study of farmer opinions on the vital issues affecting them, especially those of organizing within agriculture or forming coalitions with other groups, can only lead to one conclusion—a high degree of confusion, uncertainty, and disunity existed among farmers. There is no indication that the dwindling number of farmers had produced a greater degree of unity among those remaining. Indeed, the differences of opinion seemed as great as ever. The prospect of farmers solving some of their most serious problems through organized effort appeared as remote in the 1970s as they were in the 1890s or the 1920s. The farmer who said "we are a strange breed" may have been more accurate than he realized.

Past and Future

XIII

IN THE EARLY YEARS OF THE REPUBLIC, farmers had been a clear majority in American society. In the late nineteenth and early twentieth centuries they lost their primacy as business, industry, transportation, and other aspects of the economy grew at a more rapid rate than did agriculture. Nevertheless, farmers continued to be an important plurality of the population up to World War II. In the 1940s, however, the modernization and industrialization of agriculture made rapid gains, and within a single generation commercial farmers had become a tiny minority in American society.

In 1920 there were about 32 million Americans living on the nation's 6.5 million farms. That was 30 percent of the population. Two decades later there were still 30.5 million people on approximately the same number of farms. But beginning in the 1940s, the number of farms and farmers began to decline swiftly, and this downward trend accelerated during subsequent decades. By 1981 farmers made up only a little less than 3 percent of the population. This shift from huge majority to small minority has been one of the most fundamental changes in all of American history. It reflects the basic transformation from an agricultural society to one predominantly industrial and urban.

Several factors combined to produce this major economic and social transition. In the first place, technological and scientific developments within agriculture itself were a major agent for change. The mechanization of agriculture and the application of science in crop and livestock production greatly increased the efficiency of labor and reduced the need for numbers of people in farming. In 1978 agriculture provided only 3.3 percent of the nation's employment compared to 17.5 percent in 1940. A farmer with modern equipment in 1980 could farm five to eight times as much land as his father had farmed, and do it easier and better. The technological, chemical, and genetic revolutions that occurred in agriculture after World War

II permitted the businessman-farmer with land, capital, and management skills to develop large and highly efficient operations.

Secondly, while revolutionary changes were occurring within agriculture, off-farm job opportunities grew rapidly. American farms had always produced surplus population, but opportunity for nonfarm jobs varied in different historical periods. For example, there were few nonfarm jobs in the South that might have attracted poor southern farmers out of agriculture in the late nineteenth century, and during the Great Depression of the 1930s the stagnant industrial and urban economy could not provide employment for the surplus farm population. The great economic expansion accompanying World War II, however, and the subsequent growth of industry, transportation, finance, retail and wholesale trade, and the service industries, including government, provided huge numbers of new jobs. Since farm incomes were relatively low, millions of farmers and their children took employment in towns and cities. Despite all of the oratory about the desirability of farm life, farmers knew better than anyone else how wide the gap was between rhetoric and reality. Sons and daughters of farmers left the old homestead by the millions after 1940, more often than not with their parents' blessing.

And why? Many rural people simply did not like farm life, while millions of others saw the cities as places of opportunity. But the main reason that people left the farm can be found in relatively low incomes and standards of living. In 1948 the average income of people on farms was 67 percent of that enjoyed by the nonfarm population, in 1960 it was only 54 percent, and in 1970, 74 percent. From the early nineteenth century onward, the low return earned by most farmers compared to that enjoyed by other Americans was a major factor in causing people to leave agriculture for more profitable employment. It was no different in the period after World War II.

While many people voluntarily left the farm for better opportunities elsewhere, some were forced out of agriculture against their will. Land once farmed by sharecroppers and tenants was taken over by owners who were able to cultivate larger acreages with modern machinery. And many small farm owners were not able to earn enough income to keep going; they either went broke, or sold out to larger and more prosperous neighbors. As the amount of nonfarm inputs rose in agriculture, thereby increasing the capital requirements, hundreds of thousands of small farmers could not obtain the necessary cash or credit to continue. Furthermore, government programs after 1933 gave the most help to larger operators who had land and a substantial amount of commercial production.

With price supports and diversion payments, the larger farmers had a degree of financial security that permitted them to buy the land of their less fortunate neighbors. Most large commercial operators in the 1960s and 1970s were cultivating lands that had once been farmed by several families. As these families gave up and moved away, the buildings were usually torn down, the fences ripped out, and the fields enlarged to meet the requirements of the new machines. This process not only changed farming, it changed the entire face of rural America.

The most important factor responsible for the rapid decline in the number of farms after World War II, however, was the almost constant cost-price squeeze on farmers. Many producers could not continue farming, because they were unable to adjust to that age-old disadvantage of having no control over the price of the products they sold or purchased. As agriculture became more commercialized, farmers did not possess that most basic right in business of being able to set the price of their product to cover costs and also leave something for labor and return on investment. The cost-price squeeze was at the heart of the drive by farmers to get bigger, to increase efficiency, and to lower unit costs. When prices did not provide an adequate return on a farmer's current operation, the pressure was to increase volume and cut costs in order to make up for the lower margin of profit. Unit costs could be reduced by increasing output in relation to inputs of labor and capital. Thus, the wheat or corn farmer lowered his cost per bushel by spreading machinery expenses over more acres and producing more bushels per acre. The dairyman improved his position by milking more cows and increasing production per cow. For most farm operations, the needed efficiency required larger units. It was the greater volume at lower unit costs which gave those larger farmers more income.

The alternative to increased size and efficiency was for farmers to be able to set their prices at a level which would have given them a satisfactory income with less volume. However, because of their strong feelings of independence, their differing and often conflicting interests, and their lack of political power, they were unable to achieve such a goal either through their own efforts or through government. Some farmers supported the idea that government should guarantee prices equal to the cost of production plus a fair profit, but Congress refused to provide that degree of support. The various federal programs helped hundreds of thousands of farmers, but most of the commercial producers placed their main reliance on growth and cost cutting.

Most small and poorer farmers could not join the process of modernization. Having neither the land nor the capital required for mid-twentieth-century commercial farming, they rapidly dropped out of agriculture after 1945. Their land was taken over by the bigger operators. Critics of this development argued that government should have provided special credit and price policies for smaller farmers, and perhaps even placed restrictions on the large producers. It was occasionally suggested that limits be placed on farm size, and that nonfarm corporations be prohibited from engaging in agricultural operations. But political support could not be mustered for proposals that would have placed limits on people's aspirations and chances for success, and would have threatened private property rights. The larger operators were especially critical of such ideas, and they were the ones who held whatever political clout farmers retained. Moreover, while there may have been social and cultural reasons to keep many small farmers on the land, that option did not make economic sense. It ignored the fundamental fact that there was a surplus of people in agriculture, especially in the South.

Assuming that the goal of private enterprise and government policy was to give people an opportunity to earn a decent living, the proposals to keep people on small farms were unacceptable. The income that could be earned from cultivating 50 to 100 acres was simply not enough in the vast majority of cases to provide modern housing, automobiles, electricity, furniture, appliances, education, and other amenities of modern living. It would have taken massive subsidies, or off-farm work, to give such farmers enough money to meet their needs. Without an infusion of outside help, to keep farmers on small farms was to relegate most of them to permanent poverty and second-class citizenship. No one knew this better than small farmers themselves. That is the reason some of them sought to get bigger, while others got part-time jobs off the farm or completely abandoned agriculture. There was no conspiracy against the small farmer, unless the general trends in the overall economy are considered conspiratorial. As America modernized, and as incomes and standards of living rose, most people wanted to join the process. The economic problems of most people left behind in the rural areas could not be solved on the farm. There was a surplus of labor in most farm regions that could not be fully or profitably employed with the land and capital available to them. Congress and the American people, however, failed to see the situation as it was. The role of government should not have been to keep people in farming, but to develop policies that would have eased their transition out of agriculture. Since little was attempted or achieved in this regard, the needed adjustment between population and land was brought about by unrestrained economic forces. This caused severe hardship for many people, especially in the poorer agricultural areas.

Many people were deeply disturbed by the declining farm population. The thought of a few hundred thousand big producers dominating agriculture in place of millions of family farmers aroused the strong emotions buried in the agrarian tradition. But, as had been true for a century or more, the smaller family farm found its strongest supporters among those who did not have to make a living on the farm. Presidents, congressmen and senators, bureaucrats, editors, writers, teachers, and poets—people who would have starved if they had been forced to farm for a living—continued to lead the song of praise for agriculture. Fewer and fewer Americans, however, were taking the rhetoric of agricultural fundamentalists very seriously.

Farmers themselves had come under the influence of a competing and even more compelling tradition. That was the idea of bigness and efficiency. Ever since the 1890s, efficiency had been a key concept in American business and government. It would have been unrealistic to assume that agriculture could have escaped the same trends. Most Americans were imbued with the theory of social and technological progress, and they usually equated big with better. In agriculture, large southern planters and western ranchers had been among the nation's most admired citizens. Who was more truly American than that hard-bitten, independent cattleman or the courtly and genteel cotton grower. Their place in American culture had been immortalized through a progression of novels and movies which provided a favorable image of their way of life. Thus when the modern farmer gained control of a thousand or two thousand, or more acres of land, he personified that basic American tradition of size and success. Some of his neighbors might be critical and envious, but most of them wished they could emulate him.

Many farmers who had not become big operators retained the hope that some day they might achieve such a status. That is one reason why there was no widespread support among farmers for limiting the size of farms. As one farmer told a group of economists and politicians early in 1980, "we do not want our opportunity to get bigger hindered by federal laws designed to keep us small, even if this means pushing someone else out of farming as we get bigger." In farming, then, where genuine competition prevailed, it became a struggle for survival. Only the fittest could survive under the economic conditions facing farmers, and they became fewer and fewer. The surprising fact was not that so many farmers went out of business in the years after World War II. The amazing thing was that so many hung on as long as they did in the face of difficult problems, inferior incomes, and relatively low standards of living.

The forces and policies operating in the early 1980s indicate that the number of farmers will continue to decline. Most conditions favor the larger operator. The big producer with a higher income benefits most from the tax and credit policies affecting farmers. Furthermore, inflation enhances the wealth of those who already own land and provides the bigger landowners a capital base to purchase even additional acres. The high cost of land and equipment for modern farming makes it very difficult for new farmers to enter the business and contributes to further consolidation and larger units. At no other time in American history was it so hard to enter farming as in the late 1970s. The old saying that in order to get a farm a young man had to inherit it or marry it was never more true. Besides the prohibitive capital and operating costs required of the beginning farmer, there were no signs that the cost-price squeeze was abating. Indeed, it was accelerating. A bushel of corn would purchase nearly 10 gallons of tractor diesel fuel in 1973 compared to less than 3 gallons in 1979. After attending several farm sales early in 1980, a family-type farmer in eastern South Dakota said that he had never seen farmers so discouraged and "blue." Predictions that farm income would be down anywhere from 20 to 30 percent in 1980, and operating costs up as much as 12 or 13 percent, made farm prospects appear bleak, he said. As it turned out, prices were better than expected but production was down. Net farm income from farming dropped from about $30.1 billion in 1979 to about $22 billion in 1980, while operating expenses rose about 15 percent. If history is any guide to future developments, the cost-price squeeze will continue to eliminate marginal producers, and many of those not so marginal.

Recognizing the special problems confronting farmers, many Americans felt deep concern about the future of the family farm. In 1979 and 1980 Secretary Bergland, national legislators, farmers and their leaders, economists, social critics, and others met, discussed, and argued over the family farm's fate. Almost everyone praised the family farm as a vitally important American institution and one that should be preserved. Pointing to the fact that family farmers produced the great majority of American agricultural products, some observers stressed that the family farm was healthy and thriving. Others were not so sure. They saw the growing number of big farms as a trend that would eventually eliminate all but a few of the largest family-type operations. Much of this discussion was academic. The fact was that very few family farmers would survive even if the great majority of farms properly fit that classification. This was because not many viable commercial farms of any kind were going to remain in the United States. By 1978 only 200,000 farms produced some two-thirds of all agricultural products; a mere 50,000 of the largest operators received about 35 percent of the total cash receipts from farming. Many families would continue to live on farms while making their living at off-farm work, but it had become clear by 1980 that not only family farmers but any kind of commercial farmers would become an ever-shrinking group in American society.

Much of the concern for the family farm was based on an erroneous image of life on the farm. While it was still a way of life, farming had become mainly a large and serious business. Many urban Americans, however, thought of the family farm as they had remembered it in their youth. Their nostalgic memories were highly selective. With warm and emotional feelings they recalled birds singing in the springtime, the earthy smell of newly plowed fields, the taste of mother's freshly baked bread, and how life seemed to be so peaceful and simple. The comparison between their rural memories and the hurried pace of urban life and the pressures of an industrial civilization favored the earlier day. They conveniently forgot the hard work, the monotonous schedules demanded in the care of livestock, the dirty labor of cleaning barns and chicken houses, the filling of kerosene lamps to fight back the dark, the worry over weather, money, and prices, and the late trips out back on frigid winter nights. Although in practice the majority of Americans had long ago abandoned farm life, the sentiments that they retained about living on the land were still deeply embedded in their thought and feelings.

Although the agrarian tradition was weakening as fewer and fewer people had any direct contact with farming, it still had an unseen and kind of mystical power over many people. The belief that farmers were good people, and that the nation would be better off with a strong agricultural component, explains much of the support farmers enjoyed in Washington. After 1933 the United States spent billions of dollars on a huge grid of agricultural aid, including price supports, credit legislation, payments for conservation, and disaster relief. Such widespread and costly assistance would not likely have been supplied without the deeply held public consensus that the economic, political, and social health of the nation depended on a thriving agriculture. The problem facing modern commercial farmers, however, was how long could they count on the national sympathy and support flowing from this Jeffersonian tradition? In all likelihood, the nation's agrarian heritage will lose most of its significance and meaning within another generation.

Many of the people who left the farms in the 1930s, 1940s, and 1950s never lost that emotional attachment to their rural roots. Some of them held on to at least part of the old family farm, giving them a direct tie to their heritage even though they may have been living in Peoria or Philadelphia. To their children, however, the farm meant little except as a prospective inheritance and a supplementary source of income. It was not that the children disliked the old family homestead, they just never had any meaningful contact with it. The grandchildren, who will enter adult life in the 1980s and 1990s, will be completely oblivious to conditions on a working farm and will have no real concept of the rural life of their grandparents. These young Americans will have about the same detached interest in what their grandparents tell them about the farm as those youths had in the early twentieth century whose ancestors related stories about Indian fighting on the frontier or about the Civil War. Although deeply held traditions die slowly, the current generation of Americans is probably the last that can stir up sympathies by talking about the virtues of farm life and the importance of preserving the family farm. This does not mean that country living will diminish in popularity. To the contrary, it will probably become more attractive as people seek quietude and open spaces. But the professional or business person or the factory worker who acquires some land, builds a home, and raises a few head of livestock as a hobby will have no real relation to farming.

Now that commercial farmers have become such a small minority in American society, how and by what means will they be able to protect their special interests? Farmers still have political influence well beyond what their numbers justify. Besides having strong friends in Congress who sincerely believe that farmers need and deserve price and income assistance, farmers can provide a crucial swing vote in some states in close elections. Moreover, farm groups have become more skillful in their use of political power. But the prospect for agricultural politics in the future is not bright. In the presidential campaign of 1980 neither major candidate paid much attention to farmers. Ronald Reagan, the winner, admitted that he did not even understand the term "parity prices," a concept that had been sacrosanct among farmers for nearly two generations.

Besides their other problems, farmers are too divided by geography, outlook, and interests. Their power is fragmented not only among commercial producers, but between large farmers and small and part-time operators. Moreover, the idea of making effective alliances or coalitions with other political groups is more hope than reality. Nor is the prospect favorable for farmers to get together, organize, and set the prices for their products at a profitable level.

Strong consumer resistance to higher food prices is a major barrier to genuine prosperity on the farm. Indeed, there is more determination to restrain food prices than to control almost any other commodity purchased by Americans.

The best hope for farmers is the increasing worldwide demand for food, and the nation's need for exports to pay for oil and other imports. American farmers have an incomparable ability to produce. If those needing food have the money to buy, and the federal government does not curb exports as it did several times in the 1970s and in 1980, the position of farmers can be improved. During 1980 farm exports hit a record of about $40 billion, despite a partial embargo. In time, of course, there may be so few farmers that they will be able to unite and negotiate prices that will adequately reflect costs, plus a fair profit. That development, however, if it does occur, is far in the future.

In the 1980s, and perhaps beyond, it seems likely that the current trends among commercial farmers will continue. Farms will get fewer and bigger, requiring increasing amounts of capital and skilled management. The traditional family farm will show further declines as larger-than-family operators and corporations produce a still greater proportion of the country's food and fiber. There will likely be widespread discussion about stopping or modifying this trend, but the development will not be checked or reversed.

Different directions in American commercial farming should not be expected, because there is no effective policy-making process in the national government to deal with structural issues in agriculture. For example, even the explicit policy of limiting irrigation farms using water from federal projects to 160 acres in order to increase the number of family units, as provided in the Reclamation Act of 1902, has never been enforced. Another example of federal indecision and inaction has been in the development of a national land policy. This has been discussed without positive results since the 1930s.

In all likelihood, the future of commercial farming in the United States will be influenced by the same forces that have brought it to its current state. These include advances in science and technology, competition for land and other resources, the cost-price squeeze, the advantages of time of entry into farming, availability of foreign markets, a patchwork of federal programs, and the talent and decisions of individual producers. Although the family farm will continue in some form for many years, the nature of this hallowed institution will change even more drastically than it did in the 1960s and 1970s. A disturbing problem facing commercial farmers, of course, is the cost and shortage of energy. In light of fuel shortages, will farmers be able to continue a type of agriculture that is so energy-intensive? Because of the importance of food, it seems unlikely that farmers would be denied the energy they need to produce food and fiber. But the price will be high. Rising energy costs will simply be another factor in the elimination of the smaller commercial farmers.

The question is not whether farmers will be a tiny minority in American society. That has been answered. The bigger issue is what effect, if any, this course of events will have on American life and society. Will the depopulation of rural America change the national character? What about the beliefs and value systems that have been associated with life on the land? Will Americans be able to establish an acceptable value system that rests on life in cities, perhaps symbolized by concrete and neon lights? How will the American character and American attitudes be affected when there is no longer a flow of country-raised people into urban centers? No answers exist for these important questions. Many Americans believe deeply that the decline in the number of farmers and the weakening of rural-based values will have an undesirable influence upon society in the United States. That aspect of agrarianism is still strong. But, as always, this attitude is more a feeling, an emotion, rather than anything based on scientific or statistical evidence. It cannot be said that the United States will be a worse or a better place in which to live with the diminution of farmers and farms. It is certain, though, that America will be different.

Some observers have compared the importance of the sharp decline in the number of farms and farmers after 1945 to the disappearance of the frontier in the 1890s. While both developments reflected fundamental changes in society, neither produced any shock or break in the progress or rhythm of American history. The excess national energies that had gone into conquering the frontier, and later into building a modern commercial agriculture, transferred smoothly to other activities, which varied from the construction of shopping malls to the exploration of outer space. Modern Americans, including commercial farmers, have accepted the science and technology that destroyed the old ways of farming.

The new, industrialized agriculture drastically changed many rural communities. Hundreds of small towns have died, and others are moving rapidly toward extinction. With them have gone schools, churches, and other rural institutions which nurtured millions of Americans who feel deeply that something basic and meaningful is being lost forever. Their ideals and value systems are strongly associated with the land, the farms, and rural institutions. Many citizens are still struggling in the early 1980s to reconcile modernized agriculture and its effect upon rural America with emotions and traditions that lie deep in preindustrial farming.

Notes

Comments on Sources

The primary and secondary sources on twentieth century American agriculture are most extensive. Some of the best information can be found in the publications of the United States Department of Agriculture. Since the mid-1930s, the Department has annually published a separate volume of agricultural statistics. Before that time statistical series were printed in the *Yearbook of Agriculture. The Yearbook of Agriculture* contains articles and special studies of great historical importance on a wide variety of subjects. They cover such topics as farm mechanization, land use, improvements in plant and animal breeding, scientific advances, and hundreds of other items. The USDA has also published many bulletins which carry the results of special studies on about every topic relating to farming in the United States. A most important source of statistics are the decennial censuses of agriculture.

Other government documents of great value have originated in Congress. The *Congressional Record* provides debates on farm issues as well as many reprinted editorials, articles, and other statements of value to researchers in agricultural history. Of greater value are the hundreds of volumes of Senate and House hearings on the scores of bills relating to agriculture, especially since the 1920s. Extensive hearings have been held on such important issues as commodity legislation, agricultural credit, the family farm, farm poverty, and other questions.

Farm periodicals and publications of farm organizations provide another rich source for agricultural history. In this study I used, among others, the *Farm Journal* (Philadelphia), *Wallace's Farmer* (Des Moines), *Capper's Farmer* (Topeka), the *Prairie Farmer* (Chicago), and the *Progressive Farmer* (Raleigh, N.C.). *Farmland News* and its predecessors under different names, published by Farmland Industries Inc., the country's largest farmer cooperative, was also most useful. The reports of the annual meetings of the Farmers Union, the American Farm Bureau Federation, and the Grange were used to trace the position of those organizations on agricultural matters. *The Nation's Agriculture*, published by the American Farm Bureau Federation, was used extensively for the period around World War II when the influence of that organization was particularly strong on agricultural policy questions.

Since the farm problem was so much in the public eye from the 1920s to the 1960s, newspapers and periodicals carried hundreds of articles and editorials relating to agriculture. Since 1954 I have maintained a clipping collection of farm materials that proved to be most useful in preparing this book. The clippings are from both metropolitan and small town newspapers. The *Des Moines Register* has carried some of the best material on agriculture and farm policies. At certain times the *New York Times* and the *Wall Street Journal* sent reporters into the farm country to get stories based on first-hand farming experience. Popular periodicals such as *Time* and *Fortune* carried articles on agriculture and farm policy. During the 1940s, when Ladd Haystead was agricultural editor for *Fortune*, that magazine carried some especially revealing articles on modern agricultural change. The best materials, however, can be found in *Agricultural History*, *The Journal of Farm Economics* and other scholarly journals.

Important books not cited in the footnotes include, Thomas S. Harding, *Two Blades of Grass: A History of Scientific Development in the United States* (Norman: University of Oklahoma Press, 1947); Harold Barger and Hans H. Landsberg, *American Agriculture, 1899-1939: A Study of Output, Employment and Productivity* (New York: National Bureau of Economic Research, 1942); and United States Department of Agriculture, *Changes in Efficiency: A Summary Report, 1966*, Statistical Bulletin 233 (Washington, 1966).

On the South see Theodore Saloutos, *Farmer Movements in the South, 1865-1933* (Berkeley: University of California Press, 1960); J. H. Street, *The New Revolution in the Cotton Economy: Mechanization and Its Consequences* (Chapel Hill: University of North Carolina Press, 1957); and John L. Fulmer, *Agricultural Progress in The Cotton Belt Since 1920* (Chapel Hill: University of North Carolina Press, 1950).

Two additional books dealing with policy which are of great importance are *Foundations of Farm Policy* (Lincoln: University of Nebraska Press, 1970) by Luther Tweeton, and Marion Clawson's *Policy Directions for U.S. Agriculture* (Baltimore: The Johns Hopkins Press, 1968). Lauren Soth, one of the keenest observers of American agriculture and farm life in the mid-twentieth century and writer for the *Des Moines Register* and *Tribune*, has written a most interesting and sensitive book entitled, *An Embarrassment of Plenty, Agriculture in Affluent America* (New York: Thomas Y. Crowell Company, 1965). Hiram M. Drache has provided a strong defense of efficiency in modern agriculture in *Tomorrow's Harvest: Thoughts and Opinions of Successful Farmers* (Danville, Ill.: Interstate Printers and Publishers, 1978). On farmer political behavior see Angus Campbell and others, *The American Voter, An Abridgement* (New York: John Wiley and Sons, Inc., 1964), chapter 13.

Chapter 1. The Vanishing Majority

1. Roy M. Robbins, *Our Landed Heritage, The Public Domain 1776-1936* (Princeton: Princeton University Press, 1942), parts I and II. See also Paul W. Gates, *The Farmer's Age, Agriculture, 1815-1816* (New York: Holt, Rinehart, and Winston, 1960), chapters 3 and 4.

2. Gates, *The Farmer's Age*, chapters 13 and 14.

3. Douglass C. North, *The Economic Growth of the United States, 1790-1860* (Englewood Cliffs, N.J.: Prentice Hall, 1961), p.189.

4. *The Essays of Ralph Waldo Emerson* (New York: Random House, 1944), p.547. Illustrated Modern Library edition.

5. Allan G. Bogue, Jerome M. Clubb, Carroll R. McKibbin, and Santa A. Traugott, "Members of the House of Representatives and the Processes of Modernization, 1789-1960," *Journal of American History*, 63 (September 1976), pp.284-85.

6. House of Representatives, *Committee on Agriculture*, House Document 91-350, 91st Cong. 2d Sess. (Washington, 1970), p.1.

7. See Fred A. Shannon, *The Farmer's Last Frontier* (New York: Farrar and Rinehart, 1945), chapter 12.

8. *Yearbook of Agriculture*, 1904 (Washington: GPO, 1905), pp. 639 and 684.

9. *Southern Cultivator*, 4 (December 1846), p.186; N. S. Hubbard, "The Relative Value of Farming Among the Occupations of Life," *Agriculture of Massachusetts*, 1872, part I, p.284.

10. H. Coleman, speaking in New York before the Munro County Agricultural Society and quoted in the *Southern Cultivator*, I (5 April 1843), p.34; quoted in Percy W. Bidwell and John I. Falconer, *History of Agriculture in the Northern United States, 1620-1860* (New York: Peter Smith, 1941), p.205.

11. *The Ohio Farmer*, 89 (14 May 1896), p.420; George E. Waring, Jr., "Life and Work of the Eastern Farmer," *Atlantic Monthly*, 39 (May 1877), p.590; and *Report of the Industrial Commission on Agriculture and Agricultural Labor*, 10 (Washington, 1901), p.151. James Bryce, *The American Commonwealth*, vol. II (New York: Macmillan Co., 1913), pp.924-25; *Prairie Farmer*, 46 (13 November 1875), p.361; Alvin Johnson, *Pioneer's Progress* (New York: The Viking Press, 1952), p.23. See Sam Rayburn's statement in the *Bonham* (Texas) *Daily Favorite*, 23 June 1916; Wickson was quoted in the *Prairie Farmer*, 51 (19 June 1880), p.193.

12. Benjamin F. Thomas, "Advantages of Rural Pursuits," in State Board of Agriculture, *Annual Report, Agriculture of Mass., 1896*, part II (Boston, 1863), pp.46-55. James M. Swank, *Notes and Comments* (Philadelphia: Allen, Lane, and Scott, 1897), pp.184-85; J. Sterling Morton to H. F. McIntosh, 2 June 1896, USDA, Secretary's Letterbook No. 17, National Archives, Record Group 16. On advances in labor productivity on the farm see C. D. Kinsman, *An Appraisal of Power Used on Farms in the United States*, USDA, Bulletin No. 1348 (July 1925), p.1.

13. Waring, "Life and Work of the Eastern Farmer," p.586.

14. Gilbert C. Fite, *An Economic History of the United States* (Boston: Houghton Mifflin, 1965), p.302.

15. Roy V. Scott, "Milton George and the Farmers' Alliance Movement," *Mississippi Valley Historical Review*, 45 (June 1958), p.93.

16. *Report of the Commission on Country Life*, 60 Cong., 2d Sess., Sen. Doc. 705 (Washington, 1909), p.5.

17. David B. Danbom, *The Resisted Revolution, Urban America and the Industrialization of Agriculture, 1900-1930* (Ames: Iowa State University Press, 1979).

18. *Report of the Commission on Country Life*, pp.13-25; and William L. Bowers, *The Country Life Movement in America, 1900-1920* (Port Washington, N.Y.: Kennikat Press, 1974).

Chapter 2. American Farms and Farmers

1. Ladd Haystead and Gilbert C. Fite, *The Agricultural Regions of the United States* (Norman: University of Oklahoma Press, 1955); *Yearbook of*

Agriculture, 1930 (Washington, 1930), p.942; *1930 Census of Agriculture*, General Report, IV (Washington, 1932); and *1950 Census of Agriculture*, General Report, II (Washington, 1952).

2. *Yearbook of Agriculture*, 1930, p.414.

3. William R. Camp, "The Organization of Agriculture in Relation to the Problem of Price Stabilization," *Journal of Political Economy*, 32 (June 1924), pp.282-314.

4. *Yearbook of Agriculture*, 1922 (Washington, 1923), p.4.

5. Joseph G. Knapp, *The Rise of American Cooperative Enterprise, 1620-1920* (Danville, Ill.: Interstate Printers & Publishers, 1969); and Gilbert C. Fite, *Beyond the Fence Rows: The History of Farmland Industries, Inc., 1929-1979* (Columbia: University of Missouri Press, 1978), chapter 1.

6. Clarence A. Wiley, *Agriculture and the Business Cycle Since 1920* (Madison: University of Wisconsin Studies in the Social Sciences and History, No. 15, 1930), p.17; *Yearbook of Agriculture*, 1923 (Washington, 1924), p.95. On the general farm depression in the early 1920s see James H. Shideler, *Farm Crisis* (Berkeley: University of California Press, 1957).

7. *Congressional Record*, 67 Cong., 1 Sess., 5 July 1921, p.3332.

8. *Yearbook of Agriculture*, 1923, pp.9-10.

9. William G. Murray, "Farm Mortgage Foreclosures in Southern Iowa, 1915-1935," *Iowa Research Bulletin*, 248 (Ames, 1938), pp.249-76.

10. Letters in the files of the Secretary of Agriculture, Agricultural Situation, 1920 and 1921. National Archives, R.G. 16.

11. *Yearbook of Agriculture*, 1923, pp.8-11.

Chapter 3. The Minority Fights Back

1. Murray R. Benedict, *Farm Policies of the United States* (New York: Twentieth Century Fund, 1953), p.189; O. M. Kile, *The Farm Bureau Through Three Decades* (Baltimore: Waverly Press, 1948), chapters 2 and 3 and pp. 64 and 94; Christiana M. Campbell, *The Farm Bureau and the New Deal* (Urbana: University of Illinois Press, 1962), chapter 1; and William G. Carleton, "Gray Silver and the Rise of the Farm Bureau," *Current History* 28 (June 1955), p.348.

2. Arthur Capper, *The Agricultural Bloc* (New York: Harcourt, Brace and Company, 1922), chapter 1; Richard Lowitt, *George W. Norris, The Persistence of a Progressive, 1913-1933* (Urbana: University of Illinois Press, 1971), p.173.

3. Capper, *The Agricultural Bloc*, pp.3-5.

4. Benedict, *Farm Policies of the United States*, p.183; see James H. Shideler's *Farm Crisis, 1919-1923* (Berkeley: University of California Press, 1957) for a full account of congressional actions to help farmers; and also Donald L. Winters, *Henry Cantwell Wallace, As Secretary of Agriculture, 1921-1924* (Urbana: University of Illinois Press, 1970).

5. *New York Times*, 3 October 1921, p.8; 23 November 1921, p.14; and 22 December 1921, p.6.

6. Gilbert C. Fite, *George N. Peek and the Fight for Farm Parity* (Norman: University of Oklahoma Press, 1954), chapters 2-6.

7. Mark Sullivan, "The Waning Influence of the Farmer," *World's Work*, 51 (April 1926), pp. 657-61.

8. Quoted in Fite, *George N. Peek*, p.175.

9. "Political Dynamite of the Coolidge Veto," *Literary Digest*, 97 (9 June 1928), pp.5-7.

10. Julien N. Friant to George N. Peek, August 21, 1928. Chester C. Davis Papers, Western Historical Manuscripts Collection, University of Missouri, Columbia.

11. *Agricultural Statistics*, 1962 (Washington, 1963) p.510; Rainer Schickele, *Agricultural Policy, Farm Programs and National Welfare* (Lincoln: University of Nebraska Press, 1954), p.144.

12. See letters, Mrs. Wellman Bruner to Secretary Arthur Hyde, 4 June 1930; Mrs. Terra McCrae to Hyde, 31 July 1931, and Mrs. M. M. Clayton to Hoover, 5 April 1932. Files of the Secretary of Agriculture, National Archives, R.G. 16.

13. Clifford B. Anderson, "The Metamorphosis of American Agrarian Idealism in the 1920's and 1930's," *Agricultural History*, 35 (October 1961), pp.182-88.

14. William D. Rowley, *M. L. Wilson and the Campaign for the Domestic Allotment* (Lincoln: University of Nebraska Press, 1970); Fite, *George N. Peek*, pp.229-33.

15. Theodore Saloutos, "Edward A. O'Neal: The Farm Bureau and the New Deal," *Current History*, 28 (June 1955), pp.356-61.

16. John L. Shover, *Corn Belt Rebellion, The Farmers' Holiday Association*, (Urbana: University of Illinois Press, 1965).

17. Edwin G. Nourse, Joseph S. Davis, and John D. Black, *Three Years of The Agricultural Adjustment Administration* (Washington: The Brookings Institution, 1937); Van L. Perkins, *Crisis In Agriculture, The Agricultural Adjustment Administration and the New Deal* (Berkeley: University of California Press, 1969).

18. Gilbert C. Fite, "Farmer Opinion and the Agricultural Adjustment Act, 1933," *Mississippi Valley Historical Review*, 48 (March 1962), p.666; Perkins, *Crisis in Agriculture*, pp.51-52.

19. Henry A. Wallace, *New Frontiers* (New York: Reynal & Hitchcock, 1934), p. 188. See also Edward L. and Frederick H. Schapsmeier, *Henry A. Wallace of Iowa: The Agrarian Years* (Ames: Iowa State University Press, 1968), and Dean Albertson, *Roosevelt's Farmer, Claude R. Wickard in the New Deal* (New York: Columbia University Press, 1955).

20. Carl T. Schmidt, *American Farmers in the World Crisis* (New York: Oxford University Press, 1941), pp. 125 and 216-18; Benedict, *Farm Policies of the United States*, pp.281-82.

21. Nourse, Davis, and Black, *Three Years of the Agricultural Adjustment Administration*, pp. 194 and 202.

22. Sidney Baldwin, *Poverty and Politics* (Chapel Hill: University of North Carolina Press, 1968); David E. Conrad, *The Forgotten Farmers, The Story of Sharecroppers in the New Deal* (Urbana: University of Illinois Press, 1965); Paul E. Mertz, *New Deal Policy and Southern Rural Poverty* (Baton Rouge: Louisiana State University Press, 1978); and Donald Holley, *Uncle Sam's Farmers, The New Deal Communities in the Lower Mississippi Valley* (Urbana: University of Illinois Press, 1975).

23. *Farm Tenancy*, Report of the President's Committee (Washington, 1937), pp. 25 and 65; *16th Census of the United States, Agriculture, 1940, General Report, III* (Washington, 1943), p.912.

24. Baldwin, *Poverty and Politics*, p.108.

25. Baldwin, *Poverty and Politics*, pp.244ff.; and Benedict, *Farm Policies of the United States*, pp.262-64.

26. Baldwin, *Poverty and Politics*, pp.383-400.

27. *New York Times*, 15 June 1940, p.10.

Chapter 4. The Quickening Pace of Agricultural Change

1. *1950 Census of Agriculture, General Report, II* (Washington, 1952), pp. 223 and 232.

2. *Yearbook of Agriculture*, 1940 (Washington, 1940), p.515.

3. *Yearbook of Agriculture*, 1931 (Washington, 1931), pp.1068-69; *Agricultural Statistics*, 1942 (Washington, 1942), p.685.

4. Walter W. Wilcox, *The Farmer in the Second World War* (Ames: Iowa State College Press, 1947), p.9.

5. "The Farmer's Future," *Newsweek*, 16 (23 December 1940), p.36; *Yearbook of Agriculture*, 1940, p.552.

6. Paul S. Taylor, "Good-By To The Homestead Farm," *Harper's*, 182 (May 1941), pp.589-97.

7. Louis Cantor, *A Prologue To The Protest Movement* (Durham, N.C.: Duke University Press, 1969), pp.64-66.

8. Temporary National Economic Committee, *Investigation of Concentration of Economic Power*, Agriculture and the National Economy, Monograph No. 23, 76 Cong., 3 Sess. (Washington, 1940), p.5; and "The 32,000,000 Farmers," *Fortune*, 21 (February 1940), pp.68-71.

9. Hazel Hendricks, "Farmers Without Farms," *Atlantic*, 166 (October 1940), pp.461-68.

10. Quoted in Cantor, *A Prologue To The Protest Movement*, p.156. The letters were written 7 June and 21 January 1939.

11. Quoted in Taylor, "Good-By To The Homestead Farm," p.593.

12. Taylor, "Good-By To The Homestead Farm," p.594.

13. USDA, *Technology On The Farm*, A Special Report By An Interbureau Committee and the Bureau of Agricultural Economics (Washington, August 1940), pp.82-93. See also the excellent articles in the 1940 *Yearbook of Agriculture* entitled *Farmers in a Changing World* (Washington, 1940) that deal with technology (pp.509-32), the surplus farm population (pp.870-86), and farm management problems (pp.489-508). John A. Hopkins deals with the effect of technology on farm employment in *Changing Technology and Employment in Agriculture*, Bureau of Agricultural Economics, USDA (Washington, 1941). Wayne D. Rasmussen has analyzed the importance of technology in his article, "The Impact of Technological Change on American Agriculture, 1862-1962," in *Journal of Economic History*, 22 (December 1962), pp.578-91.

14. "The 32,000,000 Farmers," *Fortune*, 21 (February 1940), p.194.

15. *Productivity of Agriculture, United States, 1870-1958*, USDA Technical Bulletin 1238 (Washington, 1961), p.11.

16. For a discussion of some of the poorest farmers see J. Wayne Flynt, *Dixie's Forgotten People: The South's Poor Whites* (Bloomington: Indiana University Press, 1979). Dorothea Lange and Paul S. Taylor caught the spirit of rural poverty in *An American Exodus: A Record of Human Erosion* (New York: Reynal & Hitchcock, 1939).

Chapter 5. Farmers in Wartime

1. *The Nation's Agriculture*, 16 (January 1941), pp. 2, 3, 12, and 22.
2. *Farm Journal*, 65 (June 1941), pp.13-14.
3. *The Nation's Agriculture*, 16 (January 1941), p.5; and 17 (January 1942), p.24.
4. *Farm Journal*, 65 (August 1941), p.12.
5. Walter W. Wilcox, *The Farmer In The Second World War* (Ames: Iowa State College Press, 1947), p.41.
6. *Farm Journal*, 66 (June 1942), p.7.
7. *The Nation's Agriculture*, 16 (November 1941), p.3; *Congressional Record*, 77 Cong., 2 Sess., 29 January 1942, p.854, and 15 January 1942, p.410.
8. Quoted in Wilcox, *The Farmer In The Second World War*, p.245.
9. *Congressional Record*, 77 Cong., 2 Sess., 2 October 1942, p.7739; and *The Nation's Agriculture*, 18 (January 1943), p.6.
10. USDA, Economics, Statistics, and Cooperative Service, *Farm Income Situation*, Statistical Bulletin 609 (Washington, July 1978), pp. 37 and 49; *Farm Income Situation*, August-September 1952, pp. 32 and 45.
11. Wilcox, *The Farmer In The Second World War*, pp. 98-101, 264, and 293; See also the important study by John L. Shover, *First Majority Last Minority* (DeKalb: Northern Illinois University Press, 1976), p. xiv. The Yearbook of Agriculture, 1943-1947, entitled *Science and Farming* (Washington, 1947), has excellent articles on new products, machines, insecticides, and other aspects of a changing agriculture.
12. Allen J. Matusow, *Farm Policies and Politics In The Truman Years* (Cambridge, Mass.: Harvard University Press, 1967), p.5.
13. *Ibid.*, chapters 1-3.
14. *Congressional Record*, 77 Cong., 2 Sess., 20 October 1942, pp.8488-89, and Matusow, *Farm Policies*, p.40.
15. "Greed on The Farm," *Life*, 20 (20 May 1946), p.36; "Farm Bloc On The March," *Business Week*, 2 March 1946, p.15.
16. Martin E. Shirber, "The Low-Income Farmer," *Commonweal*, 65 (30 November 1956), p.227.
17. "Why Potatoes Are Being Burned," *U.S. News and World Report*, 22 (30 May 1947), pp. 14 and 51; and Matusow, *Farm Policies*, pp.126-30. See also Senate Committee on Agriculture and Forestry, *Price Supports For Perishable Products: A Review of Experience*, Committee Print, 82 Cong., 1 Sess. (Washington, 1951).
18. Matusow, *Farm Policies*, p.138.
19. Samuel Lubell, *The Future of American Politics* (New York: Harper, 1952), p.161; and Matusow, *Farm Policies*, chapter 8.
20. Quoted in *Congress and the Nation, 1945-1964* (Washington, 1965), p.688.
21. Matusow, *Farm Policies*, p.196.
22. Reo M. Christensen, *The Brannan Plan, Farm Politics and Policy* (Ann Arbor, Mich., 1959), chapter 4.
23. USDA, Economics, Statistics and Cooperative Service, *Farm Income Situation*, Statistical Bulletin No. 609 (July 1978), p.31.

Chapter 6. Problems, Progress, and Policies in the 1950s

1. *Time*, 61 (21 February 1953), pp. 22 and 53.

2. *Daily Oklahoman* (Oklahoma City) 29 October 1953.

3. *Daily Republic* (Mitchell, S.D.), 17 October 1953.

4. *Ibid.*, 19 and 20 November, 1953; *New Republic*, 129 (9 November 1953), p.9.

5. House Document 292, 83 Cong., 2d Sess., 11 January 1954; *Daily Oklahoman*, 12 March 1954.

6. "Farmers at The Barricades," *Newsweek*, 43 (25 January 1954), p.20; *Norman* (Oklahoma) *Transcript*, 8 January and 12 April 1954.

7. Fletcher Knebel, "Do Price Supports Make Sense," *Look*, 18 (19 October 1954), pp.98-99.

8. "There's Money in Farming for Some," *U.S. News and World Report*, 37 (2 July 1954), pp. 20 and 72.

9. Richard Lewis, "The Farm Belt: Behind the Barn in Dairyland," *New Republic*, 129 (17 May 1954), pp.9-12; *Daily Republic* (Mitchell, S.D.), 17 May 1956; and Samuel Lubell, *The Revolt of the Moderates* (New York: Harper, 1956), p.159.

10. *Capper's Farmer* (Topeka, Kansas), October, 1957, p.16; *U.S. News and World Report*, 37 (13 August 1954), pp.58-59.

11. *Farm Journal*, 80 (August 1956), p.8. Russell to C. J. Meadow, 17 January 1956, Russell Papers, Agriculture IX, Box 40, Russell Library, University of Georgia.

12. *Farm Journal*, 80 (August 1956), p.8; *Daily Oklahoman*, 24 April, 27 April, and 15 May 1957; *Daily Republic*, 17 May, 21 May, and 28 July 1956; "How the Farm Belt Sees Prices and Politics," *Business Week*, 26 May 1956, pp.28-29; and *Congress and the Nation, 1945-1964* (Washington, 1965), pp.700-704 and 698 for the provisions of the Soil Bank law. For broader views of farm politics see Don F. Hadwiger, "Farmers in Politics," *Agricultural History*, 50 (January 1976), pp.156-70; and Edward L. Schapsmeier and Frederick H. Schapsmeier, "Farm Policy from FDR to Eisenhower: Southern Democrats and the Politics of Agriculture," *Agricultural History*, 53 (January 1979), pp.352-71.

13. USDA, *Statistical Bulletin* No. 561 (September 1976), p.25; *Yearbook of Agriculture*, 1960, p.150.

14. Eric Hodgins, "Farming's Chemical Age," *Fortune*, 48 (November 1953), pp.150-55.

15. Harold Beuford, "Complete Cotton Mechanization Is Here," *Progressive Farmer* 78 (March 1963), p.29.

16. Gilbert Burck, "The Magnificent Decline of U.S. Farming," *Fortune*, 51 (June 1955), pp.99ff. The scholarly literature on farm problems in the 1950s, and earlier, is every extensive. See Willard W. Cochrane, *Farm Prices: Myth and Reality* (Minneapolis: University of Minnesota Press, 1958); Theodore W. Schultz, "Agricultural Policy for What?" Murray R. Benedict, "The Supply, Price, and Income Dilemma," Dale E. Hathaway, "United States Farm Policy: An Appraisal," and G. E. Brandow, "Reflections on Farm Policy, Past and Present," are studies by leading agricultural economists which appear in the *Journal of Farm Economics*, 41 (May 1959). See also Walter W. Wilcox, *Social Responsibility in Farm Leadership, An Analysis of Farm Problems and Farm Leadership in Action*

(New York: Harper, 1956). *Can We Solve The Farm Problem? An Analysis of Federal Aid to Agriculture* (New York: Twentieth Century Fund, 1955), by Murray R. Benedict, is a comprehensive study of the main farm programs from the early 1930s to the 1950s. Edward Higbee's *Farms and Farmers In An Urban Age* (New York: Twentieth Century Fund, 1963) has some stimulating ideas. On the politics of agriculture in the 1950s see Edward L. Schapsmeier and Frederick H. Schapsmeier, *Ezra Taft Benson and the Politics of Agriculture, The Eisenhower Years, 1953-1961* (Danville, Ill.: Interstate Printers & Publishers, 1975).

17. Winifred Bryan Horner, "How Long Can We Stay on the Farm?" *Saturday Evening Post*, 228 (14 April 1945), pp.38ff.

Chapter 7. Poor Farmers, Agribusiness, and the Family Farm

1. House Document 149, 84 Cong., 1 Sess. (27 April 1955), p.1. Serial 11840.

2. *Ibid.*, p.2.

3. Edward C. Banfield, "Ten Years of The Farm Tenant Purchase Program," *Journal of Farm Economics*, 31 (August 1949), p.469.

4. T. W. Schultz, "Reflections On Poverty Within Agriculture," *Journal of Political Economy*, 58 (February 1950), pp.1-15.

5. "Should We Get Rid of Marginal Farms," *Farm Journal*, 77 (November 1963), p.20; "Fewer Farmers and Higher Prices," *Business Week*, (14 May 1955), pp.28-29.

6. *Pittsburgh Post-Gazette*, 21 April 1956; Gilbert Burck, "The Magnificent Decline of U.S. Farming," *Fortune*, 51 (June 1955), pp.99ff. *Daily Oklahoman*, 11 December 1957.

7. House Document 149, 84 Cong., 1 Sess. (27 April 1955), p.2. Serial 11840.

8. *1950 Census of Agriculture*, General Report, II, passim; *1960 Census of Agriculture*, General Report, II, (Washington, 1962) passim.

9. *Farm Journal*, 79 (22 October 1955), p.134.

10. USDA, *Contract Farming and Vertical Integration*, Agriculture Information Bulletin No. 198 (July 1958).

11. *Cooperative Consumer* (Kansas City, Mo.,) 32 (31 July 1963) p.6.

12. USDA, *Contract Farming and Vertical Integration*, Agriculture Information Bulletin No. 198 (July 1958). The question of agribusiness was discussed by John H. Davis and Ray A Goldberg in *A Concept of Agribusiness* (Boston: Division of Research, Graduate School of Business Administration, Harvard University, 1957), and by John Davis and Kenneth Hinshaw, *Farmer In A Business Suit* (New York: Simon and Schuster, 1957). A decade later a broader study was done by Ewell Paul Roy, *Exploring Agribusiness* (Danville, Ill.: Interstate Printers & Publishers, 1967). Walter Goldschmidt criticized agribusiness because of its unfavorable impact on rural communities in *As You Sow; Three Studies in Social Consequences of Agribusiness* (Montclair, N.J.: Allanheld, Osmun, 1978).

13. Quoted in USDA, *Technical Bulletin*, No. 1037 (September 1951) p.2.

14. *Legislative Policies and Programs of The National Grange for 1956* (Washington, 1956), pp.5-6; John A. Crampton, *The National Farmers Union* (Lincoln: University of Nebraska Press, 1965), p.39; *New York Times*,

10 January 1956, p.10; *Family-Sized Farms*, Hearings Before the Subcommittee on Family Farms of the Committee on Agriculture, House of Representatives, 84 Cong., 1 Sess., (October 1955).

15. *Congressional Record*, 84 Cong., 2 Sess., 25 July 1956, pp.14585-86; and 86 Cong., 2 Sess., 17 May 1960, pp.10480-81.

16. Ezra Taft Benson, *Freedom to Farm* (Garden City, N.Y.: Doubleday, 1960), p.109.

17. *Family-Sized Farms*, Hearings Before the Subcommittee on Family Farms, 84 Cong., 1 Sess., 1955, p.460.

18. *Saturday Evening Post*, 229 (15 September 1956), p.81.

19. *Farm Journal*, 80 (November 1956), p.133.

Chapter 8. Farmers Struggle to Hold Their Own

1. Quoted in *Time*, 67 (7 May 1956), p.33; "How Farm Belt Sees Prices and Politics," *Business Week*, 26 May 1956, pp.28-29.

2. National Grange, *88th Annual Session, 1954, Journal of Proceedings* (Baltimore, 1954), p.98.

3. John A. Crampton, *The National Farmers Union* (Lincoln: University of Nebraska Press, 1965), p.172.

4. Charles B. Shuman, *The Annual Address of the President of the American Farm Bureau Federation*, 11 December 1956. A pamphlet.

5. *Capper's Farmer*, 70 (February 1959), p.18; and (December 1959), p.12.

6. George S. Woodfin to Sam Rayburn, 8 February 1956. Rayburn Papers.

7. *Capper's Farmer*, 68 (September 1957), p.6.

8. Quoted in *Norman* (Okla.) *Transcript*, 10 March 1957.

9. Quoted in *Daily Republic* (Mitchell, S.D.), 6 June 1957.

10. *Daily Republic* (Mitchell, S.D.) 2 August 1957.

11. Quoted in *Ibid.*

12. *Congressional Record*, 81 Cong., 1 Sess., 19 October 1949, p.15073; 81 Cong., 2 Sess., 16 March 1950, pp.3493-94; and 10 May 1950, p.6826.

13. See the tables in Willard W. Cochrane and Mary E. Ryan, *American Farm Policy, 1948-1973* (Minneapolis: University of Minnesota Press, 1976), pp.326-327.

14. *General Farm Legislation*, Hearings before the Committee on Agriculture, House of Representatives, 86 Cong., 2 Sess., Pt. 1, p.318. *Life*, 47 (7 December 1959).

15. *Congressional Record*, 86 Cong., 1 Sess., 22 May 1959, p.8957; See also 85 Cong., 2 Sess., 27 February 1958, p.3055.

16. *Farm Journal*, 83 (August 1959, Southern ed.), p.106.

17. *Capper's Farmer*, 70 (July 1959), pp.16-17.

18. *Congressional Record*, 81 Cong., 2 Sess., 6 February 1950, p.1512.

19. Quoted in *Daily Republic*, 21 January 1957.

20. *Daily Republic*, 21 April 1958.

21. *General Farm Legislation*, Hearings before the Committee on Agriculture, House of Representatives, 86 Cong., 2 Sess., Pt. 1, p.468; and John F. Kennedy, "Special Message to Congress on Agriculture," 31 January 1962. See also G. S. Tolley, "The Administration's Score on the First Round," *Journal of Farm Economics*, 43 (December 1961), pp.1033-45, and G. E. Brandow, "The 'New' Agricultural Program—For Better or For

Worse," *Journal of Farm Economics*, 43 (November 1961), pp.1019-31. Two other excellent general studies include *The Agrarian Transition in America* (Indianapolis: Bobbs-Merrill, 1969) by Wayne C. Rohrer and Louis H. Douglas, and Earl O. Heady and others, *Roots of the Farm Problem* (Ames: Iowa State University Center for Agricultural and Economic Development, 1965).

22. Richard Strout, "Gerrymandering and Unrepresentation," *New Republic* 134 (6 February 1956), pp.20-21; John F. Kennedy, "The Shame of The States," *New York Times Magazine*, 18 May 1958, p.12; William O'Hallaren, "A Fair Share for the Cities," *The Reporter*, 21 (12 November 1959), pp.22-24. Malcolm E. Jewell, ed., *The Politics of Reapportionment* (New York: Atherton Press, 1962).

23. *Farm Journal*, 88 (August 1964), p.19.

Chapter 9. Bargaining Power for Farmers

1. Joseph G. Knapp, *The Advance of American Cooperative Enterprise, 1920-1945* (Danville, Ill.: Interstate Printers & Publishers, 1973); and Florence E. Parker, *The First 125 Years* (Chicago: Cooperative League of the U.S.A., 1956).

2. Gilbert C. Fite, *Beyond The Fence Rows, A History of Farmland Industries, Inc.* (Columbia: University of Missouri Press, 1979). The problems of getting legislation passed to increase farm marketing power have been well covered by Randall E. Torgerson in *Producer Power At The Bargaining Table* (Columbia: University of Missouri Press, 1970). USDA, *Agricultural Statistics*, 1962 (Washington, 1963), p.545.

3. *Daily Republic* (Mitchell, South Dakota), 4 September 1957.

4. National Farm Institute, *Bargaining Power For Farmers*, (Ames: Iowa State University Press, 1968).

5. *New York Times Magazine*, 16 September 1962, p.28.

6. George Brandsberg, *The Two Sides in NFO's Battle* (Ames: Iowa State University Press, 1964); Denton E. Morrison and Allan D. Steeves, "Deprivation, Discontent, and Social Movement Participation: Evidence on a Contemporary Farmer's Movement, The NFO," *Rural Sociology*, 32 (December 1967), pp.414-34; John T. Schlebecker, "The Great Holding Action: The NFO in September, 1962," *Agricultural History*, 39 (October 1969), pp.204-13; *New York Times*, 14 August 1967, p.1; NFO "Milk Dumping . . . What Did It Do," *Farm Journal*, 91 (May 1967), pp.28-29.

7. *Wall Street Journal*, 12 April 1960.

8. Committee for Economic Development, *An Adaptive Program for Agriculture* (1962), in *CED Farm Program*, Hearings before the Committee on Agriculture, House of Representatives, 87 Cong., 2 Sess., August 6-10, 28, and 29, 1962 (Washington, 1962).

9. *Ibid.*, pp. 78, 276, and 303.

10. Don F. Hadwiger, *Federal Wheat Commodity Programs* (Ames: Iowa State University Press, 1970), chapter 10.

11. *Congressional Record*, 88 Cong., 2 Sess., 7 April 1964, pp. 7125 and 7152.

12. *General Farm Program and Food Stamp Program*, Hearings before the Committee on Agriculture, House of Representatives, 91 Cong., 1 Sess., July, September, and October 1969, Ser. Q, Pt. 1, p.622. Rural poverty was a subject of much discussion and study in the 1960s. See especially, National Advisory Commission on Rural Poverty, *Rural Poverty*,

Hearings Before the National Advisory Commission on Rural Poverty, Memphis, Tenn., February 2 and 3 1967, and the final report of the Commission, entitled *Rural Poverty in the United States* (Washington, May 1968). The situation among black farmers has been described in *The Black Rural Landowners–Endangered Species* (Westport, Conn.: Greenwood Press, 1979), edited by Leo McGee and Robert Boone, and in James S. Fisher, "Negro Farm Ownership in The South," *Annals of the Association of American Geographers*, 63 (December 1973), pp.478-89.

Chapter 10. The Modern Commercial Farmer

1. *Time*, 90 (18 August 1967), p.78.
2. *Minneapolis Tribune*, 16 May 1966.
3. House of Representatives, *General Farm Legislation*, Hearings before the House Committee on Agriculture, 68 Cong., 2 Sess., Pt. 1, 15 March 1960, p.420.
4. See pertinent tables in USDA, Economics, Statistics, and Cooperatives Service, *Farm Income Situation*, Statistical Bulletin No. 609 (July 1978).
5. *Department of Agriculture and Related Agencies, Appropriations for Fiscal Year 1969*, Hearings before the Subcommittee of the Committee on Appropriations, United States Senate, 90 Cong., 2 Sess. (Washington, 1968). Pertinent pages.
6. Willard W. Cochrane and Mary E. Ryan, *American Farm Policy, 1948-1973* (Minneapolis: University of Minnesota Press, 1976), p.168.
7. Quoted in the *Wall Street Journal*, 20 October 1970, p.1.
8. *Feedstuffs*, 50 (10 July 1978), p.22.
9. *Farm Journal*, 103 (December 1979), p.43.
10. *Farmland* (Kansas City, Mo.) 34 (15 February 1967), pp.10-11.
11. Jim Hightower, *Hard Tomatoes, Hard Times*, (Cambridge, Mass.: Schenkman, 1972); *Farmland News* (Kansas City, Mo.), 39 (15 September 1972), p.1.
12. G. W. Coffman, "Corporations With Farming Operations," USDA, Economic Research Service, Agricultural Economic Report, No. 209 (Washington, 1971). See also National Farm Institute, *Corporate Farming and the Family Farm* (Ames: Iowa State University Press, 1969), and Victor Ray, *The Corporate Invasion of American Agriculture* (Denver: National Farmers Union, 1968). Center for Agricultural and Economic Development, Iowa State University, *Structural Changes in Commercial Agriculture*. Proceedings of a conference held in Chicago, 12-14 April 1965.
13. See "Big Corporations Back Out of Farming," *Farm Journal*, 95 (April 1971), p.27; *Progressive Farmer*, 84 (August 1969), p.62; *Farmland* (Kansas City, Mo.), 35 (15 February 1968), p.6, and Frank Whitsitt "Corporate Flop Triggers a Farm Boom," *Farmland News*, 40 (30 November 1973), p.17, and Whitsitt, "Corporate Farm Venture Ends in Failure," *Farmland*, 38 (15 March 1971); *Kansas Farmer* (December 1967).
14. *New York Times*, December 17, 1967, p.41.

Chapter 11. The Ever-Shrinking Minority

1. *New York Times*, 27 August 1967, Sec. 3, p.1; 23 September 1967, p.28, and 27 November 1967, p.1.

2. *Wall Street Journal*, 19 February 1970.

3. *Wall Street Journal*, 7 June 1971, p.1; Michael S. Lewis-Beck, "Agrarian Political Behavior in the United States," *American Journal of Political Science*, 21 (August 1977), pp.543-65.

4. *Chicago Tribune*, 4 January 1972.

5. *National Observer*, 6 October 1973; USDA, Economics, Statistics, and Cooperatives Service, *Farm Income Statistics*, Bulletin No. 609 (July 1978), pp. 31 and 37.

6. *Atlanta Constitution*, 9 June 1978.

7. *New York Times*, 28 June 1973, p.1; 29 June, p.53; 3 July, p.1; 6 July, p.1; and 19 October, p.61.

8. Quoted in *Mattoon Journal-Gazette* (Mattoon, Ill.), 7 September 1973; and *New Republic*, 169 (21 July 1973), p.4.

9. *New York Times*, 31 July 1975, p.20; 1 August, p.30; 9 August, p.27; and 19 August, p.1. *Chicago Tribune*, 31 August 1975.

10. *New York Times*, 16 September 1975, p.55, and 22 October 1975, p.65.

11. *Chicago Tribune*, 7 March 1976; *Farmland* (Kansas City, Mo.), 35 (31 October 1968), p.2; *Time*, 111 (12 June 1978), p.68, and 112 (3 July 1978), p.66; and *Atlanta Constitution*, 9 June 1978.

12. *Nomination of Carol Tucker Foreman*, Hearings before the Committee on Agriculture, Nutrition and Forestry, United States Senate, 95 Cong., 1 Sess., 15, 16, and 21 March 1977 (Washington, 1977). *Farmland News* (Kansas City, Mo.), 43 (30 April 1977), pp.1 and 11.

13. *Farmland News*, 46 (31 January 1980), pp. 16 and 23; *Time*, 115 (21 January 1980), p.15.

14. *Atlanta Constitution*, 3 February 1979.

15. *Farmland News*, 43 (30 September 1977); *Minneapolis Tribune*, 3 October 1977; Athens (Ga.) *Banner-Herald*, 14 January 1978; and *Atlanta Journal and Constitution*, 15 and 29 October 1977, and 10 and 11 December 1977.

16. *Atlanta Constitution*, 19 and 20 January 1978.

17. *State of American Agriculture*, Hearings before the Committee on Agriculture, Nutrition, and Forestry, United States Senate, 95 Cong., 1 Sess., Pts 1-4 (Washington, 1978).

18. *Daily Oklahoman* (Oklahoma City), 25 and 26 April 1978.

19. Athens (Ga.) *Banner-Herald*, 3 and 7 February 1979; *Atlanta Constitution*, 6 and 7 February 1979.

20. *The State of American Agriculture*, Hearings before the Committee on Agriculture, Nutrition, and Forestry, United States Senate, 96 Cong., 1 Sess., 24 January 1979 (Washington, 1979), p.35.

21. *Farmland News*, 43 (15 August 1977), p.1.

22. *Atlanta Journal and Constitution*, 2 July 1978, Sec. 3, p.1.

23. *Report of the Comptroller General . . . , Foreign Ownership of Farmland-Much Concern, Little Data* (Washington, 12 June 1978); and *Report of the Comptroller General . . . , Foreign Investment in U.S. Agricultural Land-How It Shapes Up*, (Washington, 30 July 1979). See also the *Atlanta Constitution* 8 August 1978; *Wall Street Journal*, 12 June 1979; and *Farmland News*, 45 (15 August 1979), pp.12-13.

24. There is an abundant literature on the structure of American farming. See A. Gordon Ball and Earl O. Heady, *Size, Structure, and Future of*

Farms (Ames: Iowa State University Press, 1972); Richard D. Rodefeld and other eds., *Change in Rural America, Causes, Consequences, and Alternatives* (St. Louis: C. V. Mosby Co., 1978); United States General Accounting Office, *Changing Character and Structure of American Agriculture, An Overview* (Washington, 1978); Bruce L. Gardner and James W. Richardson, *Consensus and Conflict in U.S. Agriculture, Perspectives from the National Farm Summit* (College Station: Texas A&M University Press, Texas, 1979); Lyle P. Schertz and others, *Another Revolution in U.S. Farming?* (Washington: U.S. Dept. of Agriculture, 1979); United States Senate Committee on Agriculture, Nutrition, and Forestry, *Farm Structure, A Historical Perspective on Changes in the Number and Size of Farms* (Washington, 1980); *Structure Issues of American Agriculture*, USDA, Economics, Statistics, and Cooperative Services, Agricultural Economic Report 438 (Washington, 1979); USDA, *A Time to Choose: Summary Report on the Structure of Agriculture* (Washington, January 1981) provides the final report that grew out of the public hearings held on farm structure in late 1979; Harold F. Breimyer, *Individual Freedom and the Economic Organization of Agriculture* (Urbana: University of Illinois Press, 1965); and *Can The Family Farm Survive?* Special Report 219, Agricultural Experiment Station (University of Missouri, Columbia, 1978), which contains an excellent discussion by Harold Breimyer on "Can the Family Farm Survive?—The Problem and the Issues"; and National Farm Institute, *What's Ahead For the Family Farm?* (Ames: Iowa State University Press, 1966).

On related issues of importance see Don F. Hadwiger and William P. Browne, *The New Politics of Food* (Lexington, Mass.: Lexington Books, 1978); and Harold G. Halcrow, *Food Policy for America* (New York: McGraw-Hill, 1977). A study highly critical of American farm policy is *Toward A National Food Policy* (Washington, 1976), by Joe Belden, sponsored by the Exploratory Project for Economic Alternatives.

Strongly defending the modern trends on American farms are Hiram Drache's two books, *Beyond the Furrow* (Danville, Ill.: Interstate Printers & Publishers, 1976), and *Tomorrow's Harvest* (Danville, Ill.: Interstate Printers & Publishers, 1978).

Two of the country's leading agricultural economists provide both historical and current analysis in, Don Paarlberg, *Farm and Food Policy: Issues of the 1980s* (Lincoln: University of Nebraska Press, 1980), and Willard W. Cochrane, *The Development of American Agriculture, A Historical Analysis* (Minneapolis: University of Minnesota Press, 1979).

Index